Mathematics for Social Justice

Mathematics instructors are always looking for ways to engage students in meaningful and authentic tasks that utilize mathematics. At the same time, it is crucial for a democratic society to have a citizenry who can critically discriminate between "fake" and reliable news reports involving numeracy and apply numerical literacy to local and global issues.

This book contains examples of topics linking math and social justice and addresses both goals. There is a broad range of mathematics used, including statistical methods, modeling, calculus, and basic algebra. The range of social issues is also diverse, including racial injustice, mass incarceration, income inequality, and environmental justice. There are lesson plans appropriate in many contexts: service-learning courses, quantitative literacy/reasoning courses, introductory courses, and classes for math majors. What makes this book unique and timely is that the most previous curricula linking math and social justice have been treated from a humanist perspective. This book is written by mathematicians, for mathematics students. Admittedly, it can be intimidating for instructors trained in quantitative methods to venture into the arena of social dilemmas. This volume provides encouragement, support, and a treasure trove of ideas to get you started.

The chapters in this book were originally published as a special issue of the journal, *PRIMUS: Problems, Resources, and Issues in Mathematics Undergraduate Studies.*

Catherine A. Buell is Associate Professor of Mathematics. She spends her time teaching and learning from her students at Fitchburg State University, USA, and the local prison, as well as exploring the role mathematics plays in a just society. She also enjoys time with friends, the dogs, and family.

Bonnie Shulman is Professor Emerita in the Mathematics department at Bates College, Lewiston, USA. She now lives on a farm in Greene, USA, working with home-schooled youth aged 6–12 in mathematics and science.

Mathematics for Social Justice

Edited by
Catherine A. Buell and Bonnie Shulman

LONDON AND NEW YORK

First published 2022
by Routledge
2 Park Square, Milton Park, Abingdon, Oxon, OX14 4RN

and by Routledge
605 Third Avenue, New York, NY 10158

Routledge is an imprint of the Taylor & Francis Group, an informa business

British Library Cataloguing-in-Publication Data
A catalogue record for this book is available from the British Library

ISBN13: 978-1-032-01473-9 (hbk)
ISBN13: 978-1-032-05825-2 (pbk)
ISBN13: 978-1-003-19938-0 (ebk)

DOI: 10.1201/9781003199380

Typeset in Times New Roman
by codeMantra

Publisher's Note
The publisher accepts responsibility for any inconsistencies that may have arisen during the conversion of this book from journal articles to book chapters, namely the inclusion of journal terminology.

Disclaimer
Every effort has been made to contact copyright holders for their permission to reprint material in this book. The publishers would be grateful to hear from any copyright holder who is not here acknowledged and will undertake to rectify any errors or omissions in future editions of this book.

Contents

Citation Information

The chapters in this book were originally published in the journal *PRIMUS: Problems, Resources, and Issues in Mathematics Undergraduate Studies*, volume 29, issue 3–4 (2019). When citing this material, please use the original page numbering for each article, as follows:

Introduction
An Introduction to Mathematics for Social Justice
Catherine A. Buell and Bonnie Shulman
PRIMUS, volume 29, issue 3–4 (2019) pp. 205–209

Chapter 1
Doing Social Justice: Turning Talk into Action in a Mathematics Service Learning Course
Alana Unfried and Judith Canner
PRIMUS, volume 29, issue 3–4 (2019) pp. 210–227

Chapter 2
Fighting Alternative Facts: Teaching Quantitative Reasoning with Social Issues
Mark Branson
PRIMUS, volume 29, issue 3-4 (2019) pp. 228–243

Chapter 3
Measuring Income Inequality in a General Education or Calculus Mathematics Classroom
Barbara O'Donovan and Krisan Geary
PRIMUS, volume 29, issue 3–4 (2019) pp. 244–258

Chapter 4
"There Are Different Ways You Can Be Good at Math": Quantitative Literacy, Mathematical Modeling, and Reading the World
K. Simic-Muller
PRIMUS, volume 29, issue 3–4 (2019) pp. 259–280

Chapter 5
The Brokenness of Broken Windows: An Introductory Statistics Project on Race, Policing, and Criminal Justice
Jared Warner
PRIMUS, volume 29, issue 3–4 (2019) pp. 281–299

Chapter 6
Meaningful Mathematics: A Social-Justice-Themed-Introductory Statistics Course
jenn berg, Catherine A. Buell, Danette Day, and Rhonda Evans
PRIMUS, volume 29, issue 3–4 (2019) pp. 300–311

Chapter 7
Unnatural Disasters: Two Calculus Projects for Instructors Teaching Mathematics for Social Justice
Gizem Karaali and Lily S. Khadjavi
PRIMUS, volume 29, issue 3–4 (2019) pp. 312–327

Chapter 8
Supermarkets, Highways, and Natural Gas Production: Statistics and Social Justice
John Ross and Therese Shelton
PRIMUS, volume 29, issue 3–4 (2019) pp. 328–344

Chapter 9
Mass Incarceration and Eviction Applications in Calculus: A First-Timer Approach
Kathy Hoke, Lauren Keough, and Joanna Wares
PRIMUS, volume 29, issue 3–4 (2019) pp. 345–357

Chapter 10
Math for the Benefit of Society: A New MATLAB-Based Gen-Ed Course
Paul Isihara, Edwin Townsend, Richard Ndkezi, and Kevin Tully
PRIMUS, volume 29, issue 3–4 (2019) pp. 358–374

Chapter 11
Using Graph Talks to Engage Undergraduates in Conversations Around Social Justice
Alison S. Marzocchi, Kelly Turner, and Bridget K. Druken
PRIMUS, volume 29, issue 3–4 (2019) pp. 375–395

Chapter 12
Critical Conversations on Social Justice in Undergraduate Mathematics
Nathan N. Alexander, Zeynep Teymuroglu, and Carl R. Yerger
PRIMUS, volume 29, issue 3–4 (2019) pp. 396–419

Notes on Contributors

Nathan N. Alexander earned his Ph.D. at Columbia University, USA, in Mathematics and Education. His research focuses on undergraduate mathematics education, statistical and mathematical modeling, and social networks and graphs. He is currently the James King Jr. Institute Visiting Professor in the Department of Mathematics at Morehouse College, Atlanta, USA.

jenn berg wanted to be lawyer when she grew up, and on the way to getting a good LSAT score she fell in love with mathematics. Teaching became a perfect way to express her love of language and mathematics. Since joining the faculty at Fitchburg State University, USA, she has worked on honing her teaching craft with the help of her colleagues at the college and at secondary schools in the north central Massachusetts region.

Mark Branson was trained as a geometer and received his degree from Columbia University, USA. After moving to Baltimore to work at Stevenson University, USA, he became more interested in the mathematical side of the liberal arts and the role of mathematics in creating responsible citizens. Outside of the classroom, Mark is active in the Baltimore LGBTQ+ community and is an amateur chef.

Catherine Buell is Associate Professor of Mathematics. She spends her time teaching and learning from her students at Fitchburg State University, USA, and the local prison, as well as exploring the role mathematics plays in a just society. She also enjoys time with friends, the dogs, and family.

Judith Canner graduated from North Carolina State University, Raleigh, USA, with a Ph.D. in Biomathematics Zoology in 2010. She is Associate Professor of Statistics at California State University, Monterey Bay, USA. Her current research interests include biomedical data science, mathematical biology, statistics education, and quantitative reasoning.

Danette Day works at Fitchburg State to prepare the next generation of skillful, reflective teachers and administrators. Danette is an outdoors enthusiast who spends time walking, gardening, and playing outdoors.

Through Danette's love of the outdoors and her reverence for nature, she developed a deep appreciation of the mathematics found in nature, i.e., the Golden Ratio, the Fibonacci spiral, and the mesmerizing beauty of fractals.

Bridget K. Druken is Assistant Professor of Mathematics at California State University, Fullerton, USA. She teaches mathematics content courses for future K-8 teachers of mathematics and supports single-subject mathematics credential candidates. Her research interests include how to support teacher learning through lesson study, a collaborative vehicle for improving mathematics instruction.

Rhonda Evans is Sociologist who possesses a love of teaching and learning that stems from her desire and commitment to work with students, colleagues, and community members toward achieving a socially, politically, and economically just world. It is this quest that sparked her interest in working with like-minded colleagues on the development of statistics course with a social justice theme.

Krisan Geary is Coordinator of Quantitative Support and Mathematics Instructor at Saint Michael's College, USA. She developed the Mathematics for Social Justice class at Saint Michael's, and continues to teach it periodically. She has two children, Sean and Caitlin, and a husband, Paul, with whom she loves to hike and ski.

Kathy Hoke is an Applied Mathematician who enjoys finding new ways to use old mathematics. She enjoys when her three adult sons call for a mathematics solution to a problem in their fields of religion, music, or theatre.

Paul Isihara has been Professor of Mathematics at Wheaton College, USA, since 1987. In 2011, he helped launch an applied math major with a focus on humanitarian applications of mathematics.

Gizem Karaali completed her undergraduate studies at Boğaziçi University, Istanbul, Turkey. After receiving her Ph.D. in Mathematics from the University of California, Berkeley, USA, she taught at the University of California Santa Barbara, USA, for two years. She is currently Professor of Mathematics at Pomona College where she enjoys teaching a wide variety of courses and working with many interesting people.

Lauren Keough is Assistant Professor of Mathematics at Grand Valley State University, USA. She earned her B.A. in Mathematics Education from Hofstra University, USA, and earned an M.S. and Ph.D. in Mathematics from the University of Nebraska – Lincoln, USA. She is interested in active learning, increasing the participation of under-represented groups in mathematics, and undergraduate research.

Lily S. Khadjavi has Bachelor's degree in Mathematics from Harvard University, Cambridge, USA, and a doctorate from the University of California, Berkeley, USA. On the faculty in the Mathematics Department at Loyola

Marymount University, Los Angeles, USA, Lily's interests range from algebraic number theory to statistics and the law, focusing on issues such as racial profiling.

Alison S. Marzocchi is Assistant Professor of Mathematics at California State University, Fullerton, USA. Her research focuses on improving the recruitment and retention of under-represented students in postsecondary mathematics degrees. In her free time, Alison enjoys playing volleyball; practicing yoga; cooking and eating healthy plant-based food; going to concerts; and road tripping in her convertible.

Richard Ndkezi is from Rwanda and completed an applied math major at Wheaton College, USA, in 2017. He served as Dr. Isihara's teaching assistant the first time when the General Education (GE) course was offered.

Barbara O'Donovan is Mathematics and Engineering Instructor at Saint Michael's College, USA. She is passionate about STEM education and equipping students with the critical thinking skills necessary to make a valuable contribution in our ever-changing and high-tech world. She enjoys spending time with family and friends, knitting, and almost anything that allows her to be outdoors.

John Ross is Visiting Assistant Professor of Mathematics at Southwestern University, Georgetown, USA. He earned his Ph.D. and M.A. in Mathematics at Johns Hopkins University, following a B.A. in Mathematics at St. Mary's College of Maryland, USA. His research is in geometric analysis, answering questions of special manifolds that arise under curvature flows. He enjoys overseeing undergraduate research and teaching in an inquiry-based format.

Therese Shelton is Associate Professor of Mathematics at Southwestern University, Georgetown, USA. She earned a Ph.D. in Mathematical Sciences at Clemson University, following an M.S. in Mathematical Sciences at Clemson University and a B.S. in Mathematics at Texas A&M University, USA. Her current research focuses on mathematical modeling, especially in biological systems.

Bonnie Shulman is Professor Emerita in Mathematics Department at Bates College, Lewiston, USA. She now lives on a farm in Greene, USA, working with home-schooled youth aged 6–12 in mathematics and science.

K. Simic-Muller is Mathematician and Teacher Educator at Pacific Lutheran University, Tacoma, USA. She is interested in incorporating equity, cultural responsiveness, and issues of social justice into K-16 mathematics teaching.

Zeynep Teymuroglu earned her Ph.D. at the University of Cincinnati, USA. Her Ph.D. research was in applied and computational mathematics with an emphasis on mathematical biology. Her research interests are social network analysis, mathematical modeling, and financial mathematics. She teaches Calculus, Differential Equations, and Applied Mathematics courses.

Edwin Townsend began liking math as a subject in grade school and has continued to pursue that like of math ever since and at Wheaton College, USA. He participates on the club ice hockey team and likes to surf fish especially at the Outer Banks.

Kevin Tully is from Denver, USA, and plans to do a Budapest Semester before pursuing Ph.D. studies in Pure Math. He and Edwin served as TAs the second time when the General Education (GE) course was offered at Wheaton College, USA.

Kelly Turner is in her 16th year of teaching mathematics at Loara High School in the Anaheim Union High School District (AUHSD), USA. She currently teaches Advanced Placement Calculus AB and BC and MESA (Mathematics, Engineering, Science Achievement). Kelly was the 2007 Teacher of the Year of AUHSD and most recently Loara High School's 2017 Teacher of the Year.

Alana Unfried is Assistant Professor of Statistics at California State University, Monterey Bay, USA. She received her Ph.D. in Statistics from North Carolina State University, USA, in 2016. Her research includes work on STEM education, statistics education, and factor analysis.

Joanna Wares is Associate Professor of Mathematics at the University of Richmond, USA. She earned her B.S. from the University of Michigan, Ann Arbor, USA, and then earned an M.S. and Ph.D. in Applied Mathematics and Scientific Computation from the University of Maryland, USA. After this, she worked as a postdoctoral fellow at Vanderbilt University, Nashville, USA, where, under the tutelage of Dr. Glenn Webb, she began researching population dynamics, particularly in the fields of epidemiology and oncolytic virotherapy.

Jared Warner teaches at the Department of Mathematics in the City University of New York, Guttman Community College, USA. Jared enjoys the challenge of bringing mathematics to life for his students, and the creativity this challenge draws out of him.

Carl R. Yerger earned his Ph.D. in Algorithms, Combinatorics, and Optimization from Georgia Tech, Atlanta, USA. He currently serves as Associate Professor of Mathematics and Assistant Dean for Educational Policy at Davidson College, USA. Carl's research interests are in structural graph theory, combinatorics, and sports analytics.

Introduction

Catherine A. Buell and Bonnie Shulman

Abstract: In this introduction to the *PRIMUS* Special Issue on Mathematics for Social Justice we provide a brief history of social justice in the context of undergraduate mathematics pedagogy and explain the purpose and motivation behind this movement in undergraduate education.

What we teach, how we teach, and why we teach are shaped not only by institutional and accreditation requirements but also by personal philosophy. Pedagogy has three components: the curriculum, the methodology, and social education. A social justice or equity-oriented pedagogy transcends the boundaries of race, class, and gender in any classroom. In the field of mathematics, the conversation often focuses on a mathematically-rigorous curriculum (What classes should a major take? What topics are covered in Statistics?) or methodology (problem-based learning, group work, lecture, IBL, etc.). Very rarely do we discuss how to promote equity, or the role of a mathematician (or any mathematically literate person) in a democratic society. It is often argued that these issues are extra-curricular ones, and do not "belong" in a mathematics classroom. However, there is a growing number of mathematicians who believe otherwise, for both pedagogical and ethical reasons. This special issue of *PRIMUS* provides motivation, examples, and inspiration for undergraduate mathematics professors interested in incorporating social responsibility, ethics, equity, and justice into their curriculum and pedagogy.

HISTORY

Most of the early work on mathematics and social justice began in K–12 mathematics education. The website RadicalMath.org was launched in April 2006 by Jonathan Osler who was teaching in a public high school in Brooklyn, NY. In 2007, the Radical Math Teachers group held its first national conference on math and social justice at Long Island University. Bob Moses, best known for his Civil Rights work, and author of the book *Radical Equations: Civil Rights from Mississippi to the Algebra Project* was the keynote speaker. Moses maintains that math literacy is a civil right and emphasizes how mathematics has been used as a gatekeeper to educational and personal success. Over 400 educators, parents, activists and youth attended the first annual conference Inspired by Moses, this diverse group engaged in lively wide-ranging discussions. Two of the major questions addressed were: How could social justice issues be integrated into the math curriculum as a means of enriching, and not sacrificing, mathematical content? and How do race and class affect the teaching and learning of mathematics? In 2012, the conference moved from Brooklyn to Mission High School in San Francisco, where it has remained since. Bonnie Shulman, a co-editor of this special issue, was fortunate enough to attend the conference in 2007, and in the following decade, integrated the work into her classes.

The founders of the social justice mathematics movement, led by Eric (Rico) Gutstein, were primarily K–12 teachers. However, every year there were more college educators attending workshops and events in local schools. The work of Marilyn Frankenstein on Critical Mathematics Education influenced college mathematics educators such as Jacqueline Leonard, Wanda Brooks, Joy Barnes-Johnson, and Robert Q. Berry III, who in [7] say culturally relevant and social justice instruction can offer opportunities for students to learn mathematics in ways that are deeply meaningful and influential to the development of a positive mathematics identity. We also acknowledge, however, that to be effective, these approaches require teachers to carefully reflect on, attend to, and pedagogically plan for the nuances and complexities inherent in concepts such as culture and social justice.

It is important to note that the call for social justice as part of the mathematics curriculum is an international one. For instance, in 2017, the International Community Mathematics Education and Society (MES) held its 9th annual conference in Greece (*Mathematics Education and Life at Times of Crisis*). These conferences bring together people from all over the world to advance the agenda of mathematics, equity, and justice.

The movement is growing. Recently, Catherine Buell, a co-editor of this issue, was part of a team that organized a workshop for 25 participants called *Intertwining Social Justice and Mathematics* in 2016. The team also organized a Themed Contributed Paper Session at the 2017 Joint Mathematics Meetings. These experiences provided the basis for this special issue.

SNEAK PEEK

In this issue you will find lessons and projects appropriate for all levels of mathematics. For general education courses, there is Mark Branson's "Fighting Alternative Facts: Teaching Quantitative Reasoning with Social Issues"; K. Simic-Muller's "'There are Different Ways You Can Be Good at Math': Quantitative Literacy, Mathematical Modeling, and Reading the World"; and Paul Isihara, Edwin Townsend, Richard Ndekezi, and Kevin Tully's "Math for the Benefit of Society: A New MATLAB-Based Gen-Ed Course." Alison S. Marzocchi, Kelly Turner and Bridget K. Druken's "Using Graph Talks to Engage Undergraduates in Conversations Around Social Justice" incorporates social justice into courses for pre-service teachers.

Additionally, Gizem Karaali and Lily S. Khadjavi's "Unnatural Disasters: Two Calculus Projects for Instructors Teaching Mathematics for Social Justice" and "Mass Incarceration and Eviction Applications in Calculus: A First-Timer Approach" by Kathy Hoke, Lauren Keough, and Joanna Wares work topics of social justice into a Calculus classroom. "Measuring Income Inequality in a General Education or Calculus Mathematics Classroom" by Barbara O'Donovan and Krisan Geary contains material adaptable to various levels.

Other papers provide guides for community engagement with social justice and service learning such as "The Brokenness of Broken Windows: An Introductory Statistics Project on Race, Policing, and Criminal Justice" by Jared Warner and "Doing Social Justice: Turning Talk into Action in a Mathematics Service Learning Course" by Alana Unfried and Judith Canner. Finally the collaborations, "Meaningful Mathematics: A Social-Justice-Themed-Introductory Statistics Course" by jenn berg, Catherine A. Buell, Danette Day, and Rhonda Evans and "Supermarkets, Highways, and Natural Gas Production: Statistics and Social Justice" by John Ross and Therese Shelton provide insights in the statistics classroom including activities and how to create a social justice- themed course.

Concluding the issue, Nathan N. Alexander, Zeynep Teymuroglu, and Carl R. Yerger address larger questions like how and why to create a course with social justice and how to have hard conversations in "Critical Conversations on Social Justice in Undergraduate Mathematics."

We hope this special issue is a useful contribution to current conversations, and provides motivation, justification, and concrete examples for those committed to positioning mathematics teaching and learning in service to a just world. May this issue also inspire more mathematics faculty to incorporate social justice into their personal teaching philosophy and pedagogy.

ORCID

Catherine A. Buell http://orcid.org/0000-0002-5716-2110

REFERENCES

1. Alexander, N. N., Z. Teymuroglu, and C. R. Yerger. 2019. Critical Conversations on Social Justice in Undergraduate Mathematics. *PRIMUS*. 29(3-4): 396–419.
2. berg, j., C. A. Buell, D. Danette, and R. Evans. 2019. Meaningful Mathematics: A Social-Justice-Themed-Introductory Statistics Course. *PRIMUS*. 29(3-4): 300–311.
3. Branson, M. 2019. Fighting Alternative Facts: Teaching Quantitative Reasoning with Social Issues. *PRIMUS*. 29(3-4): 228–243.
4. Hoke, K., L. Keough, and J. Wares. 2019. Mass Incarceration and Eviction Applications in Calculus: A First-Timer Approach. *PRIMUS*. 29(3-4): 345–357.
5. Isihara, P., E. Townsend, R. Ndekezi, and K. Tully. 2019. Math for the Benefit of Society: A New MATLAB-Based General Education Course. *PRIMUS*. 29(3-4): 358–374.
6. Karaali, G., and L. S. Khadjavi. 2019. Unnatural Disasters: Two Calculus Projects for Instructors Teaching Mathematics for Social Justice. *PRIMUS*. 29(3-4): 312–327.
7. Leonard, J., W. Brooks, J. Barnes-Johnson, and R. Q. Berry III. 2010. The Nuances and Complexities of Teaching Mathematics for Cultural Relevance and Social Justice. *Journal of Teacher Education*. 61(3): 261–270.
8. Marzocchi, A. S., K. Turner, and B. K. Druken. 2019. Using Graph Talks to Engage Undergraduates in Conversations Around Social Justice. *PRIMUS*. 29(3-4): 375–395.
9. Mosee, R. P. and C. E., Cobb Jr. 2001. *Radical Equations: Civil Rights from Mississippi to the Algebra Project*. Boston, MA: Beacon Press.
10. O'Donovan, B. and K. Geary. 2019. Measuring Income Inequality in a General Education or Calculus Mathematics Classroom. *PRIMUS*. 29(3-4): 244–258.
11. Ross, J. and T. Shelton. 2019. Supermarkets, Highways, and Natural Gas Production: Statistics and Social Justice. *PRIMUS*. 29(3-4): 328–344.
12. Simic-Muller, K. 2019. “There Are Different Ways You Can Be Good at Math”: Quantitative Literacy, Mathematical Modeling, and Reading the World. *PRIMUS*. 29(3-4): 259–280.

13. Unfried, A. and J. Canner. 2019. Doing Social Justice: Turning Talk into Action in a Mathematics Service Learning Course. *PRIMUS*. 29(3-4): 210–227.
14. Warner, J. 2019. The Brokenness of Broken Windows: An Introductory Statistics Project on Race, Policing, and Criminal Justice. *PRIMUS*. 29(3-4): 281–299.

Doing Social Justice: Turning Talk into Action in a Mathematics Service Learning Course

Alana Unfried and Judith Canner

Abstract: Many students experience mathematics as a neutral entity, without understanding its impact on social justice and equity. Students must understand that mathematics and statistics are powerful tools for creating social change, and that students themselves are capable to enact positive social change through their mathematical abilities. In this paper, we discuss how we have integrated both service learning and mathematical consulting into a single course to promote civic engagement by mathematics majors through professional applications. We outline methods to engage with community partners to create consulting projects for students while integrating discussions of professionalism, practice, ethics, and social justice into the classroom. We provide qualitative evidence that the integration of service learning and consulting empowers mathematics students to make a difference by doing social justice with mathematics.

1. INTRODUCTION

Mathematics is not neutral. Rather, it is a critical tool for promoting justice for all [1]. However, our students often come into mathematics classrooms believing mathematics to be irrelevant to the social issues that they face daily. Gutstein argues that:

> students need to be prepared through their mathematics education to investigate and critique injustice, and to challenge, in words and actions, oppressive structures and acts [10].

He further argues that teaching mathematics through a social justice lens is essential to combat our own inequities in mathematics education regarding underrepresented minority students, giving students the mathematical and social tools to emancipate themselves. Others have also begun to discuss more thoroughly the relationship between mathematics, statistics, and social justice [2, 13]. Lesser postulates that statistics might be considered the "grammar of social justice" in that the field of statistics gives people the tools to soundly identify and respond to social inequalities [13].

Our mathematics students deserve the following realizations: first, that mathematics and statistics are powerful tools for creating positive social change, and second, that students themselves are the ones that possess the capabilities to enact that change with their knowledge and abilities. Boalar notes that one of the best ways for students to develop the realization that their mathematical skills have applications outside of the classroom is through an open-ended, project-based learning environment [4]. In addition, students in statistics courses that incorporate projects demonstrate improved student outcomes (e.g., improved test scores, positive attitudes) [19]. A mathematics consulting course is an ideal learning environment to integrate project-based learning and civic engagement through service learning.

In order to resolve social justice inequities, students must realize that they themselves can be the ones to act and make change [15]. Rockquemore and Schaffer found that service learning led to significant attitude changes in students, specifically regarding social justice, equality of opportunities, and civic responsibility [17]. Furthermore, participating in service learning leads students to "feel engaged in societal issues and empowered by their learning to make contributions to their solution [11]." Much thought has been given to service learning in mathematics [3]. However, the possible emphasis on social justice within mathematics service learning is still emerging. Lesser considers service learning to be one of three key areas that promotes the connection between statistics (or mathematics) and social justice [12]. However, he also reminds us that a service learning course is not automatically a social justice course; rather, in order for social justice to become a thread woven into a service learning course, it is crucial to focus on the reasons for, rather than just the presence of, social inequity. Lesser also emphasizes the need to focus on long-term change rather than simply meeting short-term needs of a community partner, such as building the capacity of the community partner to continue on with a project well after the service learner leaves.

In this paper, we will discuss a mathematics consulting service learning course where students are working on a mathematics or statistics project with a community partner while discovering mathematics and social justice connections along the way. It is not solely the projects, though, that lead to the deep connections that our students make between mathematics and social justice; each additional course component enhances these connections. Students are not only performing analyses for their client, but also building key relationships with their community partners, reflecting on the work through journals, and reading about, presenting on, and discussing social justice topics and their connection to mathematics throughout the semester. Such a course exposes students to the necessity of mathematical and statistical skills in creating social equality, and also pushes students to the revelation that this work is theirs to complete.

The context of the service learning course at California State University, Monterey Bay (CSUMB), is discussed in Section 2. Section 3 covers the individual aspects of the course that lead to its success. In Section 4, evidence is given from the course that students are appropriately learning to "do" social justice, and finally, Section 5 discusses how to modify our course materials to meet your specific needs.

2. COURSE CONTEXT

A nationally recognized program [6], service learning at CSUMB is currently a part of the General Education curriculum. All undergraduate students are required to take an upper-division service learning course relevant to their major/future profession and complete a minimum of 30 hours of service to the community as a part of the course requirements. As an outcomes-based institution, every upper-division service learning course at CSUMB must encompass the following topics:

- Self and Social Awareness
- Service and Social Responsibility
- Community and Social Justice
- Multicultural Community Building/Civic Engagement

Until 2013, the CSUMB Mathematics and Statistics Department only offered an upper-division service learning course where all students assisted in mathematics classrooms in middle and high schools. Though a useful service experience for the 30–40% of mathematics majors enrolled in our teacher preparation program, the other students desired a service learning experience that would prepare them professionally for their future career. In response to student demand, we created the

Mathematics Consultants[1] upper-division service learning course. We define "mathematical consulting" as the practice of serving a client using mathematical or statistical methods as appropriate to solve a problem or answer a question. We offer the course every academic year taught by a single instructor with an average enrollment of 15 students. Although most of our students are mathematics majors who have completed, at minimum, a first course in calculus and an introductory statistics course, students from other majors with similar computational backgrounds may also take the course. In the past, all proposed projects required statistical methods. Therefore, the basic data analysis skills acquired in introductory statistics (e.g., exploratory data analysis, basic data collection, hypothesis testing, estimation) are especially important for successful consulting experiences and we often provide a "statistics refresh" early in the course (see Section 3.2.5).

3. COURSE COMPONENTS

Our service learning course weaves together many components to provide students a cohesive learning experience. For example, students acquire hands-on practice applying their skills, learning to work with others, and actually "doing" social justice through the service learning projects. However, this experience would be hollow if not for the other aspects of the course that reinforce and triangulate their field experiences. This includes time spent in class developing skills that will be useful in their service learning work, as well as assignments, discussions, and synthesizing assignments related to understanding the connections between social justice and mathematics[2]. We expand on these ideas below and discuss the community partner projects, support for student success in the community partner projects, and the integration of social justice and the classroom experience to support the connection between service and the student's future profession.

3.1. Developing the Student/Community Partner Projects

The community-based consulting projects are a large component of the course, with respect to allocated time, both by students and by the

[1]Course Catalog Description: Service learning placements in local non-profit organizations, school districts and community organizations help students deepen their understanding of mathematical and statistical principles, techniques, and methodologies for effective consulting. Students will also study how the need for mathematical and statistical analysis can influence issues of social justice and equity within the local and global community.
[2]An online Appendix includes a sample syllabus and timeline for the course.

professor. In particular, the professor begins to gather the community partner projects several weeks prior to the start of the course, as it can take considerable time to find and vet potential community partner projects. The community partners should be organizations that need mathematical consulting services to support their programs and to serve their clientele, but cannot generally afford the necessary expertise. Community partners may be local chapters of national organizations (e.g., YWCA, United Way, Boys and Girls Club), local schools, museums, common community organizations (e.g., adult schools, tutoring centers, health centers, farmers markets, legal aid services), or other non-profit organizations. Many projects tend to be more statistical in nature, as most non-profits and schools are in a position where they need evidence of their effectiveness to compete for funding, and data is the best way to demonstrate effectiveness in a grant proposal. One example from past semesters is a local community health center that was interested in analyzing patterns of patient cancellations to determine if they could improve their scheduling practices. A second example is working with a local charter school with plenty of assessment data, but no way to present it; students developed graphics and reports that could be shared with teachers, administration, and beyond to summarize the school's academic progress. Yet another group of students surveyed the local population for a community-supported fishery to determine desired price points, menu items, and more.

To begin, we ascertain the needs of the interested community partner (e.g., data analysis, data collection) and determine if there are any particular obstacles for students to serve the community partner (e.g., HIPAA or FERPA regulations, background checks, agency-specific training). Often the community partner does not know exactly what they need or what is possible; they only know they either have a lot of data or need data. In such cases, we must communicate expectations and possibilities with students and community partners to develop a valuable project achievable in a single semester. The finalization of the project may take several conversations for some community partners or a single conversation with others. Therefore, the course requires a higher workload prior to the start of semester compared with a typical course to allow time to gather and finalize community partner projects.

Once we finalize the list of potential community partners and projects, we must then determine appropriate assignments for each student. Generally, we try to pair students together in groups of two based on skill sets, interest in the project, and transportation access (a common barrier for the students to visit their community partners). We also invite all community partners to attend an in-class interview session with our students prior to project assignment so that both the students and community partners have a chance to learn about each other. The visit by the community partners to the class is very important to help ease the students' nerves about taking on a consulting project and to help them connect with the mission of the community partner. In addition, the visit provides students with a low-stakes first experience with an interview-like process. After the interview session, students and community partners both privately notify the instructor of the collaboration they would prefer during the semester, and the professor acts as the matchmaker, working to satisfy both community partner and student interests. Once we establish the assignment of the consulting groups (a.k.a. the students) to the clients (a.k.a. the community partners), the consultants must begin the process to identify the client's needs and to propose a reasonable mathematically-based approach to address those needs. The students begin this process with the first meeting.

3.1.1. The First Meeting and Staying On Track

The consultants must reach out to the clients via phone or email to introduce themselves and to set up a first meeting. Here, we take time to teach the students about professionalism in email and phone conversations to help students make a good first impression and to ease any anxiety due to their unfamiliarity with the process. The goals of the meeting are for the consultants to learn about the organization and constituents the client serves, for the consultants to understand the needs of the client, and for the consultants and clients to begin to develop a clear sense of the goals and deliverables for the project.

One of the keys to service learning success is personal interactions with the community partner [17]; therefore the first meeting is just the beginning of the process, as we advise students to schedule weekly meetings with their clients. In addition, we encourage the students to complete as much of the project on-site with the community partner as possible to increase their interactions with, and understanding of, the community partner and its constituents. As students are typically meeting with their community partners on a regular basis, they are naturally incentivized to maintain progress on the project. However, many checkpoints are built into the semester to ensure progress. Every week, each project team gives an informal verbal report on the status of their project, any issues that arose the previous week, and any questions or concerns they have moving forward. For the middle third of the semester, one class period a week is set aside as working time for project teams to meet with the instructor if they are having any particular issues. Students conduct peer reviews in class of project proposals and final reports. Community partners submit a midterm and end-of-term evaluation of each student on their project to give a more formal view of what is going well and what aspects could be improved upon for the remainder of the semester. The final report must include a community partner signature indicating that the students have met with their client to discuss the final project outcomes. Lastly, students are required to submit a time log at the end of the semester with all work hours signed off by their community partner.

3.1.2. The Project Proposal

In response to the client's needs, the consultants develop a proposal for their service project at the beginning of the service experience. The proposal is a technical document with clearly defined content such as a profile of the client, the scope of the work, necessary resources, itemized tasks, a timeline for the completion of the tasks (and the 30 hours of required service), and expected deliverables. The professor and the clients review the proposal prior to the students beginning their service projects. The goal of the proposal is to establish clear expectations for both the consultants and the clients and to provide a clear plan of action for the consultants throughout the process.

3.1.3. Capacity Building

In order to serve a client, the consultant must consider the long-term needs of the client and how to provide the client with support after their service learning experience ends [12]. Therefore, all students must identify and provide a means to build the capacity of their client to understand and, potentially in the future, to do the consulting work on their own. It is "capacity building" that distinguishes consulting as service from consulting as simply fixing a problem. The goal is that any "repeat client" proposes a new service project and does not request the same services previously provided. Examples of capacity building may include the creation of a how-to manual for data analysis, the development of software or a database to automate certain tasks, or provision of a workshop for the clients to explain how and why the consultant completed their mathematical or statistical analyses.

3.2. Promoting Successful Projects

A successful consulting experience does not develop out of thin air. Rather, we must carefully plan class time to develop appropriate student skills as various components of the projects draw near. Appropriate skills range from the the strictly mathematical challenges of consulting, such as our statistics refresh, to the social and ethical challenges of consulting, which we describe below.

3.2.1. Preparing for Difficult Dialogues

The very first day of class, students begin to prepare for the civil conversations on difficult and/or controversial topics that they will have throughout the semester. This includes the consultations with their community partners to determine partner needs and to communicate challenges and changes throughout the semester, as well as conversations they will have with peers in class on varying controversial social justice issues. To prepare for these conversations, students listen to a podcast on the LARA method [24]. The LARA (Listen, Affirm, Respond, Add) method gives people with differing perspectives a way to respond affirmatively to one another in order to encourage diverse viewpoints, yet find common ground. After listening to the podcast, each student receives a note card and writes down one social or political statement someone could make to them that would greatly offend or upset them. In small groups, students pass their card to their neighbor. Their neighbor then embodies that belief and reads the statement back, after which the original writer must respond using the LARA method. Within the first class period, students are already engaging in difficult conversations with their peers, and the groundwork has been set for civil conversation, both with clients and with peers, for the remainder of the semester.

The LARA method seems to linger in the minds of many students. Without our prompting, students often mention that it is one of the key ways that they have been able to better their communication with their client when a misunderstanding may occur. Here is an example from a student journal[3]:

> *The LARA Method is an effective tool I am able to incorporate at [service site]… Whenever [my community partner and I] have differences, I find it easier to communicate when I paraphrase what the speaker said to let them know I was listening, and to help me understand what they are looking for. I found the role playing effective, because it gave me a better idea on how to collaborate with a client who needs service.*

3.2.2. What is Service?

We want to ensure that our students understand the differences between actually serving their clients (and therefore promoting long-term positive outcomes) and simply completing tasks as temporary consultants. Therefore, several class periods are dedicated to discussions about the nature of service. We discuss questions such as:

- How does serving someone differ from helping them or fixing them?
- What is service?
- Is mathematical consulting a form of service?

[3]Student quotes from journal entries have been edited for grammar and appear throughout the article in italics.

The first item is addressed with a reading assignment and class discussion on an article [16]. The last two items are addressed with a class activity in which students are given a list of phrases such as "Working in a refugee center" and "Assisting an elderly person across a busy street." Students rank these items according to which one they believe most embodies the idea of service. The classroom is marked so that one end represents "definitely is service" and one end "definitely is not service." The students go through the items one-by-one and line up across the classroom according to what rank they gave each item. If there is disagreement, we discuss what definitions of service are leading to different viewpoints. The class writes down keywords that arise from their discussion that embody the idea of service, such as "voluntary" or "selfless." In this way, it is the students rather than the professor who define service. At the end of the class discussion, students are asked to discuss whether or not mathematical consulting can be considered an act of service.

These discussions assist in creating the correct tone for the course and attitude towards the service projects. We are doing much more than simply completing projects. We are building capacity in our partners in such a way that we can promote greater equity in society through our work.

3.2.3. Consultant Training

Most students enter this course with very little real-life experience with mathematical consulting. Therefore, we first have a discussion on what an appropriate consulting relationship looks like, starting with the first meeting [23]. Next, students perform a consulting role-play with a partner in preparation for meeting the clients for the first time as previously described in Section 3.1.1. We model the role play scenarios on the examples in [20].

3.2.4. Mathematical Consulting Ethics

One key area that must be considered when connecting social justice and mathematics and statistics is that of ethics [12]. Therefore, one class period is also dedicated to a discussion on the American Statistical Association's "Ethical Guidelines for Statistical Practice" [8]. After an introduction to the guidelines, students are given a set of "Ethical Dilemma" scenarios for which they must identify which dilemmas they may encounter, as well as how to appropriately respond based on the American Statistical Association's guidelines. This experience often becomes relevant for students later in the semester when faced with the possibilities of taking shortcuts in their data entry or analysis. Here is an example from a student journal entry:

I faced the challenge of whether to keep data I believed may have issues or to go over the data collected and be sure the information taken was accurate. I remember feeling very tired just wanting to sleep and facing the issue of needing to go home right away to finish homework due the next day. Between the two choices, I could have chosen to be unethical and provide data to my client I believed to not be accurate or to revisit the data collected and be sure my results were as accurate as possible… I know it was right of my consultant group to go over the data once more to be sure there were no mistakes. In the end, it also made me feel great overall about the accuracy of data being collected… As a mathematician, I knew then that accuracy of data and ethical practices were highly important to me, in that I did not feel comfortable sending data to the client that could have possible errors.

In a recent interview, D.J. Patil, former chief data scientist under the Obama Administration, noted the importance of teaching ethics in mathematics and statistics courses [7]. Understanding ethics is a crucial foundation for being able to combat social injustices rather than create them.

3.2.5. Statistics Refresh

Some class time, 2–3 hours, is also dedicated to a brief review of statistical practice, as all past projects have required statistical methods. Many students are asked to create surveys, so the refresh covers basic tools for survey design and implementation (e.g., Google forms), bias in question design, and how to design questions for quantitative and qualitative responses, including Likert items. The refresh also includes best practices for data entry, formatting/cleaning, and storage. Finally, the refresh overviews basic summary statistics and graphics associated with different variable types, such as bar graphs and conditional proportions for comparisons of qualitative variables between groups, and mean versus median for summary of quantitative variables.

3.3. Social Justice in the Classroom

After completing the class periods that focus on setting up students for successful consulting, most of the semester remains for more detailed discussions on the relationship between mathematics and social justice. We will focus on two key assignments that help students develop their relationship with social justice in mathematics: readings and discussions around the book *Weapons of Math Destruction* [14] and the development and delivery of student TED talks.

3.3.1. Weapons of Math Destruction

The fact that math algorithms can actually be used as a social injustice was not what I wanted to hear as a math major. While these algorithms

> *turned out to be bad, what if those [people] behind them thought it was going to be helpful?*

Perhaps the single most influential piece of the course that defines student perspectives on the relationship between mathematics and social justice, aside from the projects, is the inclusion of the book *Weapons of Math Destruction: How Big Data Increases Inequality and Threatens Democracy*, by Cathy O'Neil [14]. This book discusses how the use of mathematical models in various sectors of society has actually increased injustice, whether intentionally or not. For example, O'Neil discusses how algorithms have provoked injustices seen in the housing crisis, recidivism rates, for-profit colleges, and more. Students read one chapter a week, respond to prompts given through our online learning management system, and discuss the chapters in class.

The book chapter topics create the framework for the weekly class discussions on social justice issues. We see how mathematics directly influences society on a massive scale in many different fields. Unfortunately, this influence is not always positive, which is the focus of the book. Therefore, one word of caution is that this book tends to be one of the students' first real exposures to the possible negative consequences of the use of mathematics in society. Some students expressed the need to be overly cautious when applying mathematics to real-world problems, or exhibited a fear that if they attempt to apply mathematics to real-world problems at all, they might be causing harm (as demonstrated in the student quote at the opening of this section). To us, this is a natural reaction to an initial exposure to a concept that is contrary to their prior view of mathematics as "neutral." Perhaps, even, this is healthy.

However, our goal is not to have students finishing the semester with a fear of applying mathematics to social issues! Our goal is for them to see the power of knowledge, and in turn choose to use that power for good. We therefore make sure to include several readings to discuss the positive use of mathematics and statistics in society, as well. For instance, after O'Neil discusses the issues with crime prediction software and stop-and-frisk policies, we read and discuss the article "An Analysis of the New York City Police Department's Stop-and-Frisk Policy in the Context of Claims of Racial Bias [9]," which is an excellent example of using statistics to expose an injustice so that we can then work to resolve the injustice. We also read an article about using data for good, which highlights the work of former chief data scientist D.J. Patil [7]. Here, the readings remind students that the immense power of mathematics means that there is an immense opportunity to use math for good; that is, to better society.

3.3.2. TED Talks

Another key component of the course that deals directly with social justice is TED talks [21]. Over the course of the semester, each student is responsible for giving one 4-minute talk and leading a 15-minute discussion on their topic. The talk must be about a social justice issue of the student's choice, presented from a data-driven perspective. Topics have included for-profit prisons, medical malpractice, the Syrian refugee crisis, and mental health in college students. The variety of talks exposes students to even more areas where mathematics and statistics can be applied to promote justice, and also encourages students to present arguments with sound data to back up their claims. One student noted the following about the TED talks:

> *By learning more about the [TED topics], I am happy to now have knowledge of these topics to not just inform the public, but to also be able to analyze data on this information and come to my own conclusions with possible solutions. The interdisciplinary fields mathematics has with other subjects amazes me because I have the knowledge of specific tools which can be useful to a variety of issues and have a broader impact in society and contribute to solutions for social injustice issues.*

The TED talks become another positive example of how the use of mathematics and statistics is extremely powerful, even if simply for exposing others to societal issues.

3.4. Bringing it All Together

There are many moving parts in our course and students need to be able to synthesize each component to develop their relationship with and between social justice and their profession. Reflection is a key aspect of any service learning course as it aids students in synthesizing student experiences both in and out of the classroom [11, 18]. In our course, reflection occurs informally in classroom discussions, but it also occurs more formally through weekly journals, a final report, and a synthesis paper.

3.4.1. Weekly Journals

As previously discussed, students respond weekly to a prompt about the current book chapter or article they are reading. However, students also regularly respond to an open-ended prompt asking them to reflect on their service learning project thus far (see online Appendix C for details). It is in these writings that students really begin to demonstrate the connections they are making between their applied projects, the classroom readings and assignments, and the broader themes of mathematics and social justice.

3.4.2. The Final Report

At the end of the semester, students submit a final report that catalogs what they have completed during the semester and compares the results to their original project proposal. This report includes a summary of the deliverables they provided to their client, including capacity-building items as discussed in Section 3.1.3. Thinking ahead about the final report, one student commented:

> *If I am able to craft a strong report, ... then I will know for a fact that I am capable of tackling tough mathematical modeling projects. With that experience, I hope to work harder in implementing my knowledge of social justice in the community. With more experience, I should be able to understand the factors and parameters that lead to unfair algorithms, or analysis. I hope that my experience here at [community-partner] continues to motivate me to work with local organization and businesses to benefit community services.*

3.4.3. Synthesis Paper

The final course assignment is a synthesis paper, in which students reflect on course readings and discussions as well as their service placements. Students must fully integrate all aspects of the course into the paper by referencing their service learning experiences, course readings, and classroom discussions explicitly. Students must think back on how their experiences affected their perspective on service in their profession and think forward to how they will carry the influence of their experiences into their future profession. Although the Final Report is technical and provides little room for reflection, the Synthesis Paper allows the student to gain a broader perspective on the course and service project, untethered from the constraints of professional and technical writing.

4. DOES THIS WORK?

4.1. Student Feedback

Are students actually making appropriate connections between mathematics and social justice? We will first speak more broadly about service learning courses at CSUMB as a whole, and then we will let the students speak for themselves about their specific experiences in this course. CSUMB reported that out of 240 students in any service learning course during the 2015–16 school year:

- 97% of students feel that their attitude toward service has become more positive.

- 93% of students feel that they made a meaningful contribution to the community.
- 85% of students feel a strong sense of commitment to being involved in the community. [5]

Here are a couple of key outcomes that our mathematics service learning students achieve in this course, regarding themselves, society, and their ability to impact society.

Students realize that they can emancipate themselves from injustice (as described by [10] in Section 1):

I feel like I am at the center of a crime that is committed against poor families and that it is continuously happening without any settlement to justice. I am not only a witness of this crime but also a victim that has the need to do something about it. I hope to use my knowledge of mathematics and statistics to help dissolve this empire of injustice.

Students realize that they are genuine mathematicians:

I've never really considered myself a mathematician, but now that I have people relying on my math background to help solve their problem, I realize I have not been giving myself enough credit.

Service Learning really tests our skills in real life and that is something most schools do not offer. I know more than I thought I did. I think as I keep going, I will learn more.

Students realize the enormous role that math plays in society and social justice:

As a mathematician, I have grown from seeing mathematics and statistics as just numbers to understanding their power and how they can impact society. I understand that mathematics and statistics can be applied almost everywhere. They are useful.

Mathematicians and statisticians have enormous roles in social justice. The era of efficiency and big data is at its prime as of right now, and it is the analysts who are the gate keepers to how they function in society. This means that integrity and fairness must be with those who are developing the algorithms by which our industries will run.

I never saw math as a way of creating social justice, but now I do. It doesn't have to be a super complicated mathematical algorithm, but the small things make the biggest impact. For example, just by researching and collecting data and creating graphs and pictures, we can create social justice for others. By using math, we can paint a picture for others to see clearly. And with the knowledge researched or gained, we can use that to make a difference not only in our community, but also in society.

Students realize the significance of their talents for society:

I really feel like before the course I was ignorant about the importance of my talents and how large of an effect it has on people in our everyday lives. As I continue to read and learn more, ... I become more informed and aware of all the ways math has affected me and society... it makes me want to be involved wherever it is possible for me to contribute to the greater good of society.

4.2. Community Partner Feedback

In the last 4 years, our student consultants have served over 16 different organizations, and we have always received positive feedback on the students and their work. Community partners have used the consulting services to write quarterly reports and grant applications for state and federal funding, and to make strategic decisions in the planning of their organization. In at least two cases, students continued their service learning projects even after the course ended. Several community partners report that they are still using the how-to manuals, surveys, and databases created by the consultants, evidence of the power of capacity building.

5. DISCUSSION

The integration of service learning and mathematical consulting into a single course provides students with the opportunity to develop as professionals while promoting positive change in their community. In a service learning experience, students learn more of what being a professional mathematician is like than through completing a project that solely takes place in class. One of the major differences in a service learning project is the need for students to hone their communication skills; they must interact directly with their clients and navigate through the often messy process of mathematical problem solving. In addition, students must be set up for success as consultants through specific classroom experiences such as role-playing, training in the LARA method for difficult discussions, and general discussions of professionalism, ethics, and practice. Truly, to delineate between a regular consulting project and a service consulting project, there must be an emphasis on capacity building with the goal of serving, not just fixing or helping, the community partner. The result is a mathematical consulting service learning course that empowers students as mathematical professionals and as members of their community that are ready to enact positive social change.

5.1. Adaptation for Your Classroom

We recognize that many of our peers do not have the same infrastructure as we do to support a service learning course. If your institution is located in a small community, it may be difficult to find enough community partners and projects for a service learning consulting course. Although there may not be numerous non-profits in your area, schools and school districts are often in need of statistical analysis for reporting and grant applications. There may also be ample opportunities to serve your own campus. Most institutions have an institutional assessment department, and there are often many other departments that do not have the expertise or time available to complete a mathematical or statistical project, as discussed in [22].

If the creation of an entire course is not possible at your institution, then consider how you might integrate different components of the course within your existing curriculum. For example, it might be possible to integrate readings from *Weapons of Math Destruction* [14] into a mathematical modeling course or introductory statistics course. In a course that emphasizes quantitative literacy and/or statistical literacy, it is simple to integrate a student-created TED talk assignment on data-driven perspectives on issues in social justice. Finally, in any course in which the students engage with data, it is possible to incorporate a service learning project - even a collective project by the entire class - to provide students with experience with real, messy data and ill-defined questions about that data. The authors are happy to share our course materials with anyone interested in integrating social justice, service learning, and consulting into their mathematics curriculum.

5.2. Future Work

Most of the projects the community partners propose tend to be statistical, as many community partners need to produce information for quarterly reports, evaluate the effectiveness of services, and provide quantitative evidence for grant proposals. We suspect, though, that many of our community partners are not aware that applied mathematics may also provide solutions to their needs. Whether it is mapping an optimal pathway to preserve gasoline for a transportation service or designing a space for community use, the applications of mathematics are everywhere, but they are not always recognizable to those without mathematical training. Therefore, we will work with our community partners to discover how mathematics, and not just statistics, may provide a solution to their needs. In doing so, we hope to provide meaningful experiences for students interested in other types of applied mathematics as we empower them to make a difference by doing social justice with mathematics.

ACKNOWLEDGMENTS

We would like to thank the Service Learning Institute at CSUMB for their immense assistance in finding meaningful community partners and developing appropriate course materials.

REFERENCES

1. Ball, D.L., I. Goffney, and H. Bass. 2005. The role of mathematics instruction in building a socially just and diverse democracy. *The Mathematics Educator*. 15(1): 2–6.
2. Bergen, S. 2016. Melding data with social justice in undergraduate statistics and data science courses. In J. Engel (Ed.) *Proceedings of the Roundtable Conference of the International Association of Statistics Education*. Berlin, Germany: International Statistical Institute.
3. Black, K., K.-D. Crisman, and D. Jardine. 2013. Introductory editorial: Special issue on service-learning. *PRIMUS*. 23(6): 497–499.
4. Boaler, J. 1998. Open and closed mathematics: Student experiences and understandings. *Journal for Research in Mathematics Education* 29(1): 41–62.
5. California State University Monterey Bay. 2017. Service learning quick facts. https://csumb.edu/service/service-learning-quick-facts. Accessed 6 February 2019.
6. California State University Monterey Bay. 2017. Spotlight on service and national awards. https://csumb.edu/service/spotlight-service-national-awards. Accessed 6 February 2019.
7. Chartier, T. 2017. Data for good: Tackling trillion-dollar problems. *Math Horizons*. 24(4): 18–21.
8. Committee on Professional Ethics of the American Statistical Association. 2016. Ethical guidelines for statistical practice. http://www.amstat.org/about/ethicalguidelines.cfm. Accessed 6 February 2019.
9. Gelman, A., J. Fagan, and A. Kiss. 2007. An analysis of the New York City Police Department's stop-and-frisk policy in the context of claims of racial bias. *Journal of the American Statistical Association*. 102(479): 813–823.
10. Gutstein, E. 2006. *Reading and Writing the World with Mathematics: Toward a Pedagogy for Social Justice*. New York, NY: Taylor & Francis.
11. Hadlock, C. R., Service-learning in the mathematical sciences. *PRIMUS*. 23(6): 500–506.
12. Lesser, L. M. 2006. *Critical Values: Connecting Ethics, Service Learning, and Social Justice to Lift Our World*, pp. 2269–2271. Alexandria, VA: American Statistical Association.
13. Lesser, L. M. 2007. Critical values and transforming data: Teaching statistics with social justice. *Journal of Statistics Education*. 15(1).
14. O'Neil, C. 2016. *Weapons of Math Destruction: How Big Data Increases Inequality and Threatens Democracy*. New York, NY: Crown/Archetype.

15. Poling, L. L. and N. Naresh. 2014. Rethinking the intersection of statistics education and social justice. In K. Makar, B. de Sousa and R. Gould (Eds), *Proceedings of the Ninth International Conference on Teaching Statistics*. Flagstaff, AZ: International Statistical Institute.
16. Remen, R. N. 2017. Helping, fixing, or serving? https://www.lionsroar.com/helping-fixing-or-serving/. Accessed 6 February 2019.
17. Rockquemore, K. A. and R. H. Schaffer. 2000. Toward a theory of engagement: A cognitive mapping of service-learning experiences. *Michigan Journal of Community Service Learning*. 7: 14–25.
18. Schulteis, M. S. 2013. Serving hope: building service-learning into a non-major mathematics course to benefit the local community. *PRIMUS*. 23(6): 572–584.
19. Smith, G. 1998. Learning statistics by doing statistics. *Journal of Statistics Education* 6(3).
20. Taplin, R. 2007. Enhancing statistical education by using role-plays of consultations. *Journal of the Royal Statistical Society - Series A: Statistics in Society*. 170(2): 267–300.
21. TED Conferences, Prepare your speakers + performers. https://www.ted.com/participate/organize-a-local-tedx-event/tedx-organizer-guide/speakers-program/prepare-your-speaker. Accessed 6 February 2019.
22. Zack, M. and G. Crow. 2013. Service-learning projects developed from institutional research questions. *PRIMUS*. 23(6): 550–562.
23. Zahn, D. A. and D. J. Isenberg. 1982. *Non-statistical aspects of statistical consulting. Technical report*, Florida State University, Tallahassee, FL.
24. Zomorodi, M. 2014. Your Facebook friend said something racist. now what? http://www.wnyc.org/story/your-facebook-friend-said-something-racist-now-what/. Accessed 6 February 2019.

Fighting Alternative Facts: Teaching Quantitative Reasoning with Social Issues

Mark Branson

Abstract: Mathematics has a unique and powerful role to play in the teaching of social justice issues. There is substantial quantitative evidence for social injustice, but many citizens lack the quantitative skills to understand that evidence. A course in quantitative literacy is a unique opportunity to provide this quantitative understanding to a wide range of students in a general education context. Quantitative literacy skills provide citizens with the tools they need to critically analyze misinformation and make good decisions about civic issues.

1. INTRODUCTION

In recent years, media pundits and political thinkers have begun to confront the political consequences of poor data literacy. A variety of consequences of the modern age, including diverse media sources of varying quality, increasing ideological division, and the rise of populist politicians has brought this issue to the forefront of political discussions: where do individuals get their information, how do they analyze the validity of that information, and what action should they take on the basis of that information.

Concurrently with this, the mathematics community has been grappling with the pedagogical question of how we teach mathematics to students outside of the STEM fields. At my university, students who did not need calculus or statistics for their major were traditionally tracked into MATH 137: College Algebra as their terminal math course. However, this course was not designed as a terminal math course for non-STEM majors. It focuses on developing the abstract reasoning needed to complete Precalculus and Calculus. Non-STEM students frequently complained that the course was irrelevant and saw it as a hoop that they needed to jump through, rather than a worthy investigation of interesting knowledge. Our goal in designing a new course for non-STEM majors was based on the suggestion of Packer: to "make each year of mathematics instruction worthwhile in itself, not just preparation for the next mathematics course" [14]. In considering what material would make mathematics instruction "worthwhile" for these students, we chose to design a course in quantitative reasoning, which encouraged students to see algebraic skills through the lens of applications.

The course was designed with a series of bi-weekly student projects. The goal of these projects was to permit students to explore the topics of the course in more depth than allowed by lecture and homework. We wanted students to see that mathematics was about more than just producing calculations; that they could connect those calculations to real-world problems and use them to develop solutions. Initially these problems were largely apolitical. For example, a project on growth of functions asked students to explore linear and exponentially growing zombie populations. However, as the course coordinator, I felt such projects did the students a disservice. Although the projects may have made the material more interesting to the students, it still lacked any inherent applicability. The events of the 2016 presidential campaign led me to consider whether we would be better served by teaching students about the mathematics involved in understanding social justice.

Much of the study of teaching social justice in mathematics has been aimed at providing economically disadvantaged and minority students with the tools they need to understand inequity and make change [5, 9, 10]. The idea of connecting the oppressed with the tools they need to fight against their oppression arises from the work of Freire [7] and the culturally relevant pedagogy of Ladson-Billings [11]. Although ignorance of social issues can hurt oppressed groups through the creation of learned powerlessness, this ignorance is even more dangerous in the hands of a majority group. As Frankenstein states, "The ruling class can more effectively keep people oppressed when these people cannot break through the numerical lies and obfuscations thrown at them daily" [6]. The students participating in a private liberal arts university classroom are either already a member of the ruling class or positioned to become a part of that class. Therefore, this classroom presents an opportunity to perform a different kind of intervention - to educate these privileged students about the inequities that surround them, and to encourage them to make change.

Arguments against social justice reforms in the United States frequently take one of two forms:

1. The problem does not exist: e.g., *police officers do not engage in racial profiling, predatory lending is an acceptable business practice.*
2. The problem cannot be solved through legislative or social reform: e.g., *gun violence is inevitable, improved educational outcomes for low-income students can only be achieved through increased efforts by those students*

Support for these arguments is increasingly provided in the form of spurious data, ironically dubbed *alternative facts*. The term *alternative facts* was originally used to discredit reliably validated quantitative data [15], but has since come to signify spurious information used to justify ideological arguments. Frequently, this data is easily discredited through basic research. Due to the perceived neutrality of quantitative data, the mathematics community, and especially the quantitative literacy classroom, occupy an ideal space in education to counter these arguments. Mathematics education can provide students with the tools to counter alternative facts with reliable quantitative data, and to check the veracity of the quantitative data that they encounter in the media.

2. SOCIAL JUSTICE IN QUANTITATIVE LITERACY

At my university, Quantitative Literacy courses (MATH 133, 134, and 135) serve as general education courses for students who do not require other mathematics for their major. MATH 135, Introduction to Mathematical Reasoning, is a one-semester course covering a variety of quantitative reasoning topics: geometry, linear and exponential models, financial mathematics, data representations, probability, and statistics, using a traditional quantitative literacy text [4]. MATH 133 and MATH 134 form a two-semester sequence, offered to students who do not place into college-level mathematics, which combines the material of MATH 135 with supporting material. This course is designed to allow students who would otherwise spend a semester on non-credit-bearing courses to proceed immediately into college-level coursework. These courses serve a wide swath of the campus community, generally students with weaker mathematics backgrounds. Prior to 2014–15, when the course was developed, these students primarily took College Algebra for their general education mathematics credit.

The general education program is part of the broader set of six goals that graduates are expected to meet by the completion of their time at the university. The quantitative literacy requirement is generally associated with the first goal, Intellectual Development, which reads:

> The …graduate will use inquiry and analysis, critical and creative thinking, scientific reasoning, and quantitative skills to gather and evaluate evidence, to define and solve problems facing his or her communities, the nation, and the world, and to demonstrate an appreciation for the nature and value of the fine arts. [16]

However, the program also includes in Goal 3: Self, Societies, and the Natural World, the language:

> The …graduate will consider self, others, diverse societies and cultures, and the physical and natural worlds, while engaging with world problems, both contemporary and enduring. [16]

Traditionally, the social sciences and humanities have been seen as the primary pathway to expose students to these types of issues. STEM courses, especially at the lower levels, have been more concerned with encouraging students to think critically and exposing them to the content of the courses. Engagement with world problems, when it does come into STEM courses, is either confined to specific disciplines, such as Environmental Science, or is specifically concerned with finding scientific solutions to these problems.

Mathematics has not classically been a part of this dialogue at all. Although a considerable amount of research has been carried out on the culture of the mathematics classroom [1–3, 14], most of that work has focused on how mathematics educators communicate the values of the profession - values such as rationalism and objectism; control and progress; openness and mystery in the framework of Bishop [2]. Although the value of *progress* suggests that mathematics educators perceive and/or present mathematics as a tool for change, that presentation has largely been of mathematics as a tool for technological and scientific advancement. Although most mathematicians are perfectly willing to acknowledge the role that mathematics plays in analyzing the world's problems, such as climate change, they are often reluctant to bring those mathematical results into the classroom. The political aspect of mathematics education has largely focused on the need to improve mathematics education for underserved minorities, rather than the actual political implications of mathematical results.

Due to the perceived apoliticity of mathematics, especially pure mathematics, instructors have been reluctant to introduce subjects perceived as political into mathematics. If one can illustrate the notion of a confidence interval using a classroom poll about whether students prefer cats or dogs, why ask instead how they feel about legalized abortion?

The perception in the field has largely been that such examples distract students from the actual mathematical information and waste class time more valuably spent on mathematics.

This argument, however, is actually detrimental to the teaching of quantitative literacy. An inherent part of quantitative literacy is the understanding of how quantitative arguments relate to issues in the world. Gutstein refers to this as the ability to "read the world with mathematics" [9], and it is a critical ability for citizens in a democracy. Although specific departments may be able to focus on specific goals, such as training mathematicians, scientists, or businesspeople, the broader goal of the university is to train educated citizens. Democratic societies depend on an educated populace to make good decisions. Quantitative literacy courses are an opportunity to provide this ability to "read the world with mathematics," and thus to become better citizens, to students outside of the STEM disciplines.

Finally, awareness is increasing across the community that social justice must be a part of the mathematics education we provide. In 2016, the National Council of Supervisors of Mathematics and TODOS: Mathematics for All, released a joint statement committed to social justice in both delivery of the mathematics curriculum and in the curriculum overall [12]. More recently, organizations such as Campus Reform have begun to target mathematics and mathematics education faculty, especially faculty of color [8], to perform work in the areas of increasing access to mathematics education for diverse populations. Finally, the community is awakening to the reality that our traditional values, such as rationalism, are now newly perceived as political by parts of the populace. As anti-scientific opinions and beliefs invade the political discourse, mathematicians have little choice but to become a part of the political conversation.

3. STRATEGIES FOR TEACHING SOCIAL JUSTICE IN QUANTITATIVE LITERACY

Like Frankenstein in [5] and [6], MATH 133, 134, & 135 teach social justice through group project work. As the course is taught entirely by part-time faculty, the course was designed to be easily portable between faculty and to use familiar pedagogy. Therefore, the primary instructional technique remains a traditional lecture. Although most of the classroom time is spent in faculty lecture, time is allotted in the schedule for groups to work on bi-weekly projects. These projects are scaffolded inquiry-based exercises designed to be easy to deliver for instructors who are new to the course or have significant time commitments outside of this course.

Instructors have independence in how they manage this group work, with most preferring to allow students to self-select groups unless this becomes a problem. They also provide students with varied amount of classtime to work on group projects.

In each project, students are required to:

1. Obtain data from a real-world source.
2. Synthesize that data into a mathematical format.
3. Apply one or more concepts/techniques learned in class.
4. Reflect on the consequences of their mathematical conclusions.

Although these are common goals for mathematics exercises, students are not generally accustomed to applying these techniques critically to issues that they encounter in news sources or social media. The projects actively encourage students to make these connections. Real-world data is rarely as clean or simple as the data encountered in textbook problems, so this encourages them to develop skills in processing rich data sources and extracting important data. As these data may not be in a readily usable form, the students then need to process it before they can perform mathematical operations on them. Finally, the crucial step of reflection encourages students to process the connection between the mathematical facts and their own model of social reality.

Currently, the projects (five or six per semester, depending on schedule) are evenly split between social justice topics, such as those listed below, and less serious topics such as zombie movies (linear versus exponential growth), painting a building (geometry), and buying a car (financial mathematics). The course initially focused almost exclusively on nonsocial justice topics, which have gradually been added in as new projects are created.

A sample project is included below. Appliance and furniture rental establishments are a common part of the landscape in Baltimore city and the lower-income portions of Baltimore county. As our students come from a mix of lower and middle income neighborhoods, both urban and suburban, their level of familiarity with these establishments will vary. For some students, this project will be their first encounter with such an establishment, whereas for others they will be a recognizable part of their communities. This provides an opportunity for inter-cultural dialog between students of different backgrounds.

MATH 135

Project #3: Rent-to-Own

Today were going to study interest rates. Many of you may already be familiar with rent-to-own establishments which offer expensive consumer goods such as electronics and furniture at low monthly rates ... after a period one or two years, the item is yours. It sounds like a deal, until you do the math. These companies are actually selling you the product on credit—just like it would be if you bought it with a credit card—but at much higher interest rates.

1. Go to the website for [a corporation which operates furniture & appliance rental stores in Baltimore] and enter your zip code to see local prices. Now choose a selection of 10 products you would be interested in buying for your home—try to find a mix of items: furniture, electronics, and appliances.
2. For each item, write down the weekly price, how many payments are required to own the item at that price, and the total amount you would pay (this is the weekly price times the number of payments).
3. Now, find the item on another website where you might buy it from. Make sure that the items are exactly the same, and note the price and the website where you found that price.
4. Divide the price from the other website by the number of payments you would make to [the corporation]. This is the amount you would be paying if there were no interest.
5. Recall from class that the amount of money paid A on a loan of P dollars at an annual interest rate of r percent, compounded n times per year for t years is

$$A = P\left(1 + \frac{r}{n}\right)^{nt}$$

   This means that if we want to take the amount we paid for the item (A), the amount the item is worth (P), the number of payments per year ($n = 52$, since we make one payment every week), and the number of years we pay for it (t, the number of payments divided by 52), we can solve this equation for the interest rate r we are paying. We get

$$r = 52 * \left(\frac{A}{P}\right)^{\frac{1}{52t}} - 52$$

   Use this formula to compute the interest rate you are being charged on each of the ten items.

6. A typical student credit card will have an interest rate around 20%. Using the first formula above

$$A = P\left(1 + \frac{r}{n}\right)^{nt}$$

determine how much you would pay for each item if you put it on a credit card and paid it off in the same amount of time as the [corporation] "deal."

7. Explain in a paragraph why someone may choose to use [the corporation] instead of buying it from the website where you found it listed. How would members of your group purchase these items if they needed them?
8. Using the internet, find a newspaper article about predatory lending practices. Write a short summary of the article discussing the kind of lending, why it is predatory, and who the target of the practice is.

The data collection in this project is not challenging, but it highlights the differences between the places where people without access to credit and/or banking are able to shop. Students are required to find an article about predatory lending practices and discuss how these practices occur, to encourage deeper reflection on the causes and motivations behind the injustice.

Other projects currently used in the courses ask students to calculate crime rates and stop rates in New York City during the stop-and-frisk era, analyzing bloc voting in the Israeli parliament, the consequences and ethics of home HIV testing, and reading polls on the Iranian nuclear deal in 2015. These topics are discussed in more detail in the Appendix. In each, students are asked to delve more deeply in the topic at hand, exploring the mathematical evidence and the social consequences of that evidence. Topics are chosen to illustrate specific mathematical concepts which tie to the text (the project above links to a unit on investment, credit, and banking) and to a specific topic in social justice, but are also chosen for mathematical incontrovertibility. Although students may argue the ethics or morality of charging the poor interest rates in excess of 50%, the mathematical reality that the poor are charged interest rates in excess of 50% is impossible to argue. Students learn that although individuals can have differing opinions, those opinions should be based in fact, and that those facts should be mathematically supportable.

4. STUDENT REACTIONS

Student feedback about the social justice projects was largely negative, which is consistent with most of the feedback in the course. Although some of this negativity could come from preexisting student perceptions of quantitative literacy, it also seems that some of it was related to the way the projects were framed/delivered. A number of student comments on course evaluations indicated that they felt the projects were "busy work" or "absolutely pointless." One student asserted that the projects "were more tests of my understanding of Excel or properly researching and citing sources or writing a movie plot than what was actually being taught," indicating that the student had failed to see the connection between doing mathematics calculation and using other skills to communicate the results of those calculations. Another student declared that:

> every project was confusing and had nothing to do with math, We were talking about things I should be learning in an English class or a science class. Stick to the subject, instead of trying to mix an already difficult subject with others

Clearly there are challenges in the way that the material is being framed, and how instructor's show that it is relevant to the study of quantitative literacy.

During Spring 2018, I conducted several classroom observations of the faculty teaching this course. I observed that the faculty were providing limited guidance and limited discussion of the material before the projects. Students were engaging in very limited reflection on the consequences and implications of the projects. To aid the faculty in having these discussions, I plan to develop discussion guides for each project so that the faculty can engage students in thinking about the broader implications of the projects. I also plan to increase the amount and importance of the reflection piece of each project, in the hope that students will engage in a more thoughtful consideration of the ramifications of the quantitative results. I plan to assess these changes by further teaching observations and evaluation of student comments. I would also like to begin gathering data on how student perception of social justice topics changed over the course of the semester, so that we can evaluate how much of the negative reaction is a result of prior opinions about quantitative topics and how much is a result of the material covered in the course.

Interestingly, none of the student comments discussed the political nature of the projects. No instructor has ever registered a student complaint regarding the political nature of the projects. This may be a function of our campus population, which is largely liberal, or a well-recognized (but strictly anecdotal) tendency by our student body not to challenge authority figures.

5. FUTURE GOALS

Based on the assessment in the previous section, one of the first challenges that needs to be met is better training of part-time faculty. I hope to use course meetings in the fall to organize discussions of the project material and how faculty can engage students in discussion of the topics. Faculty observations have shown that instructors are focusing primarily on the mechanics of calculations, rather than reflection on the consequences of those calculations. Professional development for the faculty will need to focus on tools for initiating discussion in the classroom, as well as organizing the course material to allow time for these discussions.

In addition to teaching students to "read the world with mathematics," Gutstein [9] establishes a second goal of teaching them to "write the world with mathematics," which he uses to describe the process of creating social change with mathematical arguments. In his classroom, consisting of middle-school students, he finds this goal to be considerably more challenging, due to resistance from parents and the limited agency of schoolchildren. With college students, such a goal is considerably easier to achieve. In future projects, students will be encouraged to envision how they could take action on the topics in question, if they feel that action is justified. Gutstein argues that quantitative literacy goes beyond understanding the quantitative information in the world to" creating a pedagogy of questioning," [9] where students are not content with an unjust status quo. True quantitative literacy cannot exist solely in the classroom, so neither can quantitative literacy for social justice.

The challenge in such a teaching goal is avoiding instructor bias; students must be encouraged to act out of their own beliefs, rather than a desire to earn a grade or instructor approval. Many faculty in the humanities and social sciences already have extensive experience in this area, and I plan to reach out to colleagues in that area to learn the techniques that they use in their classrooms.

Another future goal of this initiative is to begin assessing social engagement in the course. Although the course is already assessed to measure student progress in quantitative literacy, no assessment data is being collected on their social justice experiences. Again, as with the goal above, this assessment must be done carefully to avoid bias. Research questions of interest include whether students change their perception of social issues, improve their ability to discern quantitative facts, and process quantitative information into social conclusions.

Finally, new projects will be introduced. In the interest of synergy with the first goal listed above, these projects will focus on the university or the Baltimore area, where students may be able to exercise more agency over the problems for which they find quantitative evidence. Possible topics include food and banking deserts (geometry), adjunct pay and cost of living (financial mathematics), Baltimore city politics (voting), and wealth inequality (financial mathematics). These projects will gradually be phased in to replace current projects.

6. CONCLUSION

Mathematicians have a duty to produce an educated populace, capable of making informed decisions using quantitative information. Traditional pedagogies, which teach mathematics completely divorced from real-world issues, do not give students the skills to extract accurate quantitative information from the wealth of data available to them, or to process that data into meaningful information. Newer pedagogies frequently avoid culturally and socially relevant problems to avoid causing controversy. This creates the impression in the mind of the student that mathematics is not a part of the broader social movements around them, and that quantitative information is something to be consumed passively rather than analyzed actively.

The passive absorption of information, which may be misleading or inaccurate, is a tremendous threat to our democracy. I believe that each of us, as educators, must do our part to fight this threat. As mathematics educators, we must make a stand for rigorous, rational, and factual use of information throughout public discourse. To do nothing only supports those who seek to use information in misleading and counterfactual ways.

APPENDIX

APPENDIX A: ADDITIONAL PROJECTS

MATH 134

Project 2: Stop and Frisk

The Stop and Frisk program was a controversial program implemented in New York City to reduce crime rates. We're going to graph some data from the New York City Police Department (NYPD) and evaluate the effectiveness of this program in reducing crime. The program spurred protests during the late 2000s and early 2010s because of allegations that stops were racially motivated.

1. Using the data from the NYCLU, create the following graphs in a program like Microsoft Excel. Note that some of these numbers are not given directly in the paper, but must be calculated from the data which you are given. You can choose the type of graph, but you should pick a graph which is appropriate for representing this data.
   (a) The number of people stopped by the police each year
   (b) The number of people who were stopped and found to be completely innocent

(c) The number of people who were stopped and not found completely innocent
(d) The percentage of people stopped from each ethnic group (black, Latino, white, and other)

2. You have also been given a spreadsheet containing data from the NYPD on major felony crime rates. Although stop and frisk arrests were frequently for minor crimes, police theorized that reducing these minor crimes would reduce the number of major crimes as well. Plot the number of each crime as well as the total number of crimes taking place in the city each year.
3. Because the population of a city is always changing, crime rates are generally calculated as a ratio of crimes per 100,000 people. You can find the crime rate per 100,000 people by taking

$$\frac{\text{number of crimes}}{\text{population}} * 100,000$$

Find the population of New York City (be sure to cite your source) for each year between 2000 and 2015 and use Excel to calculate the number of crimes per 100,000 people. Explain why this is a better way to measure crime rate than just by number of crimes.

(1) Similarly, calculate the number of police stops each year per 100,000 people. Graph both this and the crime rates from the previous problems. Does the crime rate go down when there are more police stops? What about when there are fewer police stops? Do you think that Stop and Frisk reduced major crime rates in NYC? Explain your answer.
(2) Find data on the percentage of the population in NYC that identify as black, Latino, white, or other (be sure to cite your sources). How does this data explain why people could be upset about the percentages stopped from each ethnic group?
(3) An article in The Atlantic—(warning: contains brief profanity during transcripts of police stops) discusses the controversy of racial disparities in Stop and Frisk which led to the reduction in the number of stops seen in 2012-2015. If the program were effective at reducing crime, do you think it would be acceptable for the NYC government to continue with Stop and Frisk? Are there ways that you could improve the program? Are there other ways that these large number of police stops could affect crime rates or community interactions with the police? You can use your own experiences and research to support these points, but be sure to be respectful of other members of your group. We all have different experiences and opinions, but that does not preclude us from

having civil discussions. Write at least 300 words on this topic, and make sure your answer is well-structured and readable. Be sure to cite any sources.

MATH 134

Project #5: Reading about Polls

When reading about the results of a poll in the media, very little mathematical information is generally mentioned. To find out about the methodology, sample size, margins of error, etc., you frequently need to refer to the poll report. Read the attached poll report, which asked Americans about their opinions on the Iranian nuclear deal in 2015, and answer the following questions.

1. What was the reason that the poll was given?
2. Read through the questions that were asked at the end of the poll. Do any of them seem biased to you (like theyre trying to get a particular answer, or influence people one way or the other)? Explain why or why not.
3. What was the overall sample size for the poll? Notice that many of the graphs and results are based on those who have heard about the agreement. What was the sample size for people who have heard about the agreement?
4. Calculate the margin of error for the overall sample size, and the smaller sample size for people who have heard about the agreement. On page 7, they say 58% of people believe that good diplomacy is the best path to peace—which margin of error would apply to that calculation? What about the assertion on page 5 that 42% believe the relationship between U.S. and Iran will remain about the same?
5. Although the overall sample sizes are large, the sample sizes for different subgroups are smaller. Find the number of Republicans who responded to the survey AND had heard about the agreement, and use it to compute a margin of error. Then give a confidence interval for the statement 10% of Republicans who are familiar with the deal believe the relationship with Iran will improve (from the table on page 5).
6. Find the following percentages:
   (a) Percentage of 50-64 year olds who have heard about the agreement and have confidence in the US and international agencies monitoring of Irans compliance.
   (b) Percentage of Democrats who have heard a lot about the agreement.
   (c) Percentage of college graduates who believe that diplomacy is the best path to peace.

7. Can you find any values that might be misleading if they were reported without margin of error? Hint: Look for values slightly above or below 50%, or close to the opposing valueadding/subtracting the margin of error could switch the result.
8. Find at least three news articles about this issue that present differing viewpoints, which reference this particular poll (Hint: search for articles published on or right after 7/21/2015, when the article was published). How did these different articles use this information to support their viewpoint? Do you think they were ethical to use the information this way? How could such use of statistics lead people to mistrust statistical information? Your results should be formatted in a neat paper. Be sure to cite all sources of data to avoid plagiarism! Also, you should cite the specific locations in the poll report where you found the requested data, when you answer the questions. You can divide the work between your group members however you wish, but all group members should do equal work and it should be clearly indicated which group members completed which parts.

MATH 135

Project #4: Medical Testing

Medical testing has transformed the way our society approaches public health. Rather than focusing on preventative measures like vaccination and sanitation, it allows us to target treatment and prevent diseases from spreading at the source. For this project, you will be researching and analyzing a medical test.

1. The attached article discusses the OraQuick Rapid HIV test, which is used to provide home testing for HIV. You will need to use this article and data you research for yourself.
2. Research the test and the condition it tests for. You should talk about the testing procedure, how difficult it is, how expensive it is, and how effective it is. In particular, find the false positive rate, the false negative rate, and the prevalence of the condition in the United States or Maryland (hint: some of this information can be found using Table 2 on page 15 of the article).
3. Using the data you found in the previous part, compute the specificity, sensitivity, PPV, and NPV of the test. Explain how this information tells us how good the test is.
4. Discuss the consequences of using this test. What are the possible consequences of a false-positive result? Are there other ways to confirm/refute the result? What about the consequences of a false-negative result?

5. Find (and cite) an article which discusses the OraQuick Rapid HIV test. Discuss how the mathematical accuracy of the test is explained in the article. Do you think it accurately reflects the mathematical results which you found in part 3? Discuss the ethical consequences of misleading data in an article like this. Your results should be formatted in a neat paper. Be sure to cite all sources of data to avoid plagiarism! You can divide the work between your group members however you wish, but all group members should do equal work and it should be clearly indicated which group members completed which parts.

REFERENCES

1. Bishop, A. 1988. Mathematics education in its cultural context. *Educational Studies in Mathematics*. 19(2): 179–191.
2. Bishop, A. 1991. *Mathematics Enculturation: A Cultural Perspective on Mathematics Education*. Norwell, MA: Kluwer.
3. Bishop, A., G. Fitzsimmons, W. T. Seah, and P. Clarkson. 1999. Values in mathematics education: Making values teaching explicit in the mathematics classroom. In *Combined Annual Meeting of the Australian Association for Research in Education and the New Zealand Association for Research in Education*. Deakin, ACT, Australia: Australian Association for Research in Education. https://www.aare.edu.au/publications/aare-conference-papers/show/2350/values-in-mathematics-education-making-values-teaching-explicit-in-the-mathematics-classroom. Accessed 7 February 2019.
4. Crauder, B., B. Evans, J. Johnson, and A. Noell. 2015. *Quantitative Literacy: Thinking Between the Lines*. New York, NY: Freeman.
5. Frankenstein, M. 1990. Incorporating race, gender, and class issues into a critical mathematical literacy curriculum. *Journal of Negro Education*. 59(3): 336–347
6. Frankenstein, M. 2014. A different third R: Radical math. *Radical Teacher*. 100: 77–82.
7. Freire, P. 1974. *Pedagogy of the Oppressed*. New York, NY: Seabury.
8. Gutiérrez, R. 2017. Why mathematics (education) was late to the backlash party: The need for a revolution. *Journal of Urban Mathematics Education*. 10(2): 8–24
9. Gutstein, E. 2005. *Reading and Writing the World with Mathematics: Toward a Pedagogy of Social Justice*. London, UK: Routledge.
10. Gutstein, E. and B. Peterson. 2013. *Rethinking Mathematics: Teaching Social Justice by the Numbers*. Milwaukee, WI: Rethinking Schools.
11. Ladson-Billings, G. 1995. Toward a theory of culturally relevant pedagogy. *American Education Research Journal*. 32(3): 465–491
12. NCSM, TODOS. 2016. Mathematics education through the lens of social justice: Acknowledgement, actions, and accountability. http://www.todos-math.

org/assets/docs2016/2016Enews/3.pospaper16_wtodos_8pp.pdf. Accessed 7 February 2019.

13. Packer, A. 2003. What mathematics should "everyone" be able to do? In L. Steen (Ed), *Mathematics and Democracy: The Case for Quantitative Literacy*, pp. 33–42. Princeton, NJ: Woodrow Wilson National Foundation.
14. Seah, W. T., A. Andersson, A. Bishop, and P. Clarkson. 2016. What would the mathematics curriculum look like if values were the focus? *For the Learning of Mathematics*. 36(1): 14–20
15. Sinderbrand, R. 2017. How Kellyane Conway ushered in the era of 'alternative facts.' https://www.washingtonpost.com/news/the-fix/wp/2017/01/22/how-kellyanne-conway-ushered-in-the-era-of-alternative-facts/?utm_term=.3aa0c433b00b. Accessed 7 February 2019.
16. Stevenson University. 2017. The SEE student learning goals and outcomes. http://stevenson.smartcatalogiq.com/en/2017-2018/Undergraduate-Catalog/Academic-Information/Copy-of-The-SEE-Student-Learning-Goals-and-Outcomes. Accessed 7 February 2019.

Measuring Income Inequality in a General Education or Calculus Mathematics Classroom

Barbara O'Donovan and Krisan Geary

Abstract: Income inequality is a central social justice concern, and hence excellent motivation for real-world applications in mathematics classrooms at every level. We describe the *Mathematics for Social Justice* course at Saint Michael's College, giving a specific example of one of the typical social justice projects for the course, and showing how projects can be adapted to other courses such as Calculus. The projects described focus on the Gini coefficient, a commonly used measure of income inequality. The original lesson used the trapezoid rule and Microsoft Excel to estimate the Gini coefficient for a country, while the project developed for a Calculus I course uses Maple to fit a power function to data and then integration to calculate the Gini coefficient. We also include readings on the Gini coefficient's role in policy formulation and advocacy.

1. INTRODUCTION

Saint Michael's College, a small Catholic liberal arts institution in Vermont with approximately 2000 undergraduate students, has a long history of dedication to social justice issues. The college was founded by the Society of Saint Edmund, and the centrality of social justice arises from core Edmundite values that embrace the connectedness of the world and are expressed throughout the college through education and reflection. Our students strongly align with these social justice priorities, with

Color versions of one or more of the figures in the article can be found online at www.tandfonline.com/upri.

approximately 70% of our students participating in related volunteer efforts.

When need arose in 2011 for a new liberal arts mathematics course, a course with social justice at its core was a natural outgrowth of the vision of our institution and values of our student body. The resulting four-credit *Mathematics for Social Justice* course at the general education level uses mathematical tools such as relative change, financial math, graphical displays, probability, voting theory and methods of apportionment to explore poverty, racial profiling, predatory lending, equitable political representation, unemployment, and gambling. The majority of the course consists of activities that have been developed around topics for which up-to-date data are available from sources such as the American Community Survey, U.S. Department of Health and Human Services, U.S. Bureau of Labor Statistics, and U.S. Department of Justice. Additional units are drawn from Math in Society [4] an online textbook available as an Open Educational Resource. *Mathematics for Social Justice* was initially offered in the Spring 2013 semester, and has been offered each semester thereafter. Most sections are capped at 25 students, but every other year we offer one smaller section capped at 15 students designed to support students with learning challenges in mathematics.

An important structural attribute of the course is that the topical units and their underlying mathematical content are relatively independent, so that students who may struggle with one unit may start fresh and work toward greater success with the next. Also, in response to feedback from employers and recent graduates advocating for more exposure to creating, manipulating, analyzing, and using spreadsheets, the course is rich in spreadsheet instruction and usage. Each unit is explored and investigated using Microsoft Excel, so that students develop a familiarity with spreadsheets and are able to use them in other courses and contexts. Throughout the course students learn how to effectively organize a spreadsheet, how to create a variety of graphs, how to use both explicit and recursive formulas, how to sort and filter data, and how to extend patterns to answer questions.

2. INCOME INEQUALITY AND THE GINI COEFFICIENT

In recent decades, the relationship between income inequality and health has received significant attention in public health writings. A brief literature search reveals many articles linking income inequality to public health in general, and life expectancy in particular [2, 3, 5]. It has been shown that at specific threshold income levels, increases in Gross National Product per capita are no longer associated with life expectancy gains. Thus, in affluent countries, income inequality is a more accurate predictor of life

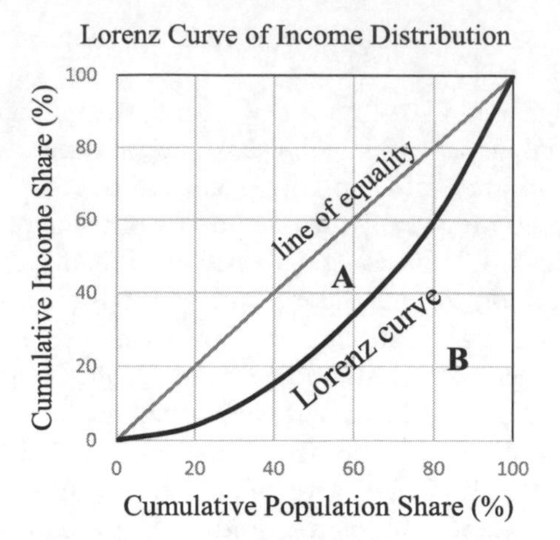

Figure 1. Lorenz curve of income distribution.

expectancy, with citizens of more equal countries exhibiting longer lifespans, on average [3]. There are numerous alternative measures of income inequality: Atkinson index, coefficient of variation, decile ratios, Thiel generalized entropy index, Kakwani progressivity index, proportion of total income earned, the Palma, Robin Hood index, and Sen poverty measure. However, the Gini coefficient is by far the most popular indicator of income inequality. Furthermore, the choice of income inequality measure does not affect the conclusion that higher levels of income inequality in a population are linked to shorter life expectancy [3, 5]. According to De Bogli, income inequality has an independent and more powerful effect on life expectancy at birth than per capita income and educational attainment. In his article, he uses this to argue that countries should enact policies that minimize income inequality and promote population health [2].

The study of how income inequality is measured by the Gini coefficient and the social ramifications of a high level of inequality make up one of the units in *Mathematics for Social Justice.* We share this unit here, as it exemplifies the components of this course. Furthermore, this unit may readily be adapted to other courses; we show how we have incorporated this material into one of our calculus courses. The unit begins with an introduction to the Gini coefficient. The Gini coefficient is defined based on the Lorenz curve which plots the cumulative percentage of the population on the horizontal axis against the cumulative percentage of total income earned by the bottom $x\%$ of the population on the vertical axis. The area below this Lorenz curve is labeled B in Figure 1. The extent to which the curve dips below the line $y = x$ indicates the degree of inequality in the distribution. The diagonal line is called the line of equality. The Gini coefficient is defined as $A/(A + B)$, the ratio of the area between the Lorenz curve and the line of equality, labeled A in Figure 1, divided by the area under the line of equality ($A + B = 0.5$ since the area of the chart is one). A Gini coefficient of zero indicates perfect equality where the bottom 20% of the population have 20% of the income, bottom 60% have 60% of the income, etc. A Gini coefficient of one indicates perfect inequality, where the top 1% has 100% of the income.

The Gini coefficient calculation results in a single summary statistic, which is very appealing and straightforward. Other attractive features of the Gini coefficient are the anonymity of high and low earners, independence of economy size, and independence of population size. However, the simplicity of the measure can lead to oversights. If used too broadly, it can also be somewhat reductionist and can overgeneralize a distribution. For example, it can be difficult to interpret the Gini coefficient because the same number can result from a variety of income distributions, caused by such things as an aging population or a country accepting a large proportion of immigrants, etc. [9].

Nevertheless, the Gini coefficient is a relative measure and can be used to compare income distribution in a variety of ways: over time, among different regions within a country, across diverse countries, etc. In order to make meaningful comparisons using the Gini coefficient, different populations need to be measured consistently. In the United States, U.S. Census Bureau data [8] is used to compute the Gini coefficient. The definition of income used by the U.S. Census Bureau is total money income of a household: wages, salaries, interest, dividends, alimony payments, child support, Social Security payments, and other cash transfers. This does not include food stamps, Medicare, or other non-cash benefits, and capital gains are not included in the census, so they are not included as income.

The World Bank [11] and Organization for Economic Cooperation and Development [6] also report Gini coefficients. Both organizations consider different information to calculate income, and thus, arrive at different values for the Gini coefficient than the U.S. Census Bureau; however, they use the same procedures and definitions for all countries their organizations consider, so that the integrity of the relative comparison is maintained. Additionally, the World Bank uses the Gini index, which is the Gini coefficient multiplied by 100.

As the Gini coefficient is such a widely used measure, and most importantly, as it is mathematically accessible, it has proved to be an especially apt topic for inclusion in a general mathematics course and for export to calculus as well. The social justice goal of these lessons is for students to understand how income inequality in the U.S. compares to other countries, appreciate the impact that income inequality has on societies, and recognize that income inequality in the U.S. is increasing. The mathematics topics that are addressed are analyzing current world data in context, estimating the area under a curve, and creating and interpreting meaningful graphical displays.

3. ACTIVITIES USING THE GINI COEFFICIENT

There are a series of activities in *Mathematics for Social Justice* that are spread over multiple class meetings. At the start of the unit, students watch a video titled "Wealth Inequality in America, Perception vs. Reality" [12] which offers a visual representation of the extent to which income is unevenly distributed in the U.S. This video kicks off the unit by introducing the issue of income inequality, showing how it has changed dramatically since the 1970s, and highlighting that it is more pronounced than people realize. This captures student interest and prompts some discussion. The first activity (see Appendix) then walks students through calculating the Gini coefficient for the United States step-by-step using World Bank data. The class, using income data for quintiles, plots and connects the data using a smooth curve to estimate the Lorenz curve. Using trapezoids, students then estimate the area under the Lorenz curve for each quintile. Students sum the trapezoidal areas to estimate the area under the Lorenz curve, labeled B in Figure 1. Next, knowing the total area under the curve is 0.5, the area between the Lorenz curve and the line of equality can be estimated using $A = 0.5 - B$. Thus, the quotient $A/(A + B)$ gives an estimate of the Gini coefficient. After students calculate the Gini coefficient for the U.S. using a table and a calculator, the class generalizes the computation process by creating an Excel spreadsheet to verify the answer and create a tool to calculate the Gini coefficient more easily in repeated calculations. Using the Excel spreadsheet as a template, smaller groups work together to edit the template to estimate the Gini coefficient for Brazil and Indonesia. Next, students compare their estimates from Excel to the Gini indices published by the World Bank and discuss their accuracy. Once the mathematical results are calculated and compared, the class discusses possible reasons why income distribution varies in different countries.

For homework, students examine specified internet websites to investigate income inequality in the U.S. By exploring the interactive websites, students learn that salaries have not kept up with inflation, learn the influence that gender and race have on income, learn that income inequality largely results from tax policy and regulations, and consider different ways to address growing income inequality. They also visit a website where they explore state-by-state income inequality information, which leads students to discover how their home state fares in comparison to others.

Next, attention turns to the changes in income distribution in the United States over five decades with another in-class activity. Students are given U.S. Census Bureau income distribution data for each decade from 1970 to 2010 in quintiles with which they create a single graph displaying the five decades of data with the line of equality and reflect on the trend that is exhibited.

Finally, following these activities the class is shown a TED talk by Richard Wilkinson titled "How Economic Inequality Harms Societies" [11] in which he discusses the relationship between income inequality and a variety of social problems. Students are given a worksheet to fill out as they watch the video. The worksheet draws their attention to public health issues as they relate to gross national income and income inequality, the data sources, how different countries narrow the gap between rich and poor, possible causes of the negative effects of income inequality, and suggestions to reduce income inequality. The students take notes on the worksheet during the video and then work with their groups to help each other fully answer all the questions afterward. Finally, there is a discussion among the whole class with each group contributing to the conversation.

The exploration of income inequality using the Gini coefficient proved highly adaptable and was easily incorporated in a standard Calculus I class. The concept of income inequality, the Lorenz curve, and calculation of the Gini coefficient is presented as a project after students have learned about integration. Before beginning the project, students are shown the same video as students in *Mathematics for Social Justice*, "Wealth Inequality in America, Perception vs. Reality" [12] and it has the same effect of prompting a preliminary discussion. At this point in the semester students have completed nine projects, are familiar with the software program Maple, and the applied nature of the projects. Like other projects in the class, the income distribution project is largely self-guided and students work in well-established groups to complete it. For calculus projects, students are given 30 to 45 minutes to work on projects in class where they can familiarize themselves with the project and ask clarifying questions. If additional time is needed, the project is completed outside of class. The project starts by introducing the Lorenz curve and draws parallels between Lorenz curves and power functions. Next students establish the relationship between the Lorenz curve and the Gini coefficient. Students then fit power curves to U.S. Census Bureau income distribution data for each decade from 1970 to 2010 and use integration to estimate the Gini coefficient. Next, they create a plot of the Lorenz curves and the line of equality. Finally, students reflect on the changes in income distribution in the U.S. over time. The project was created by merging and adapting Guided Project 19: Distribution of Wealth [1] and Applied Project: The Gini Index [7] and is included in the Appendix.

4. CONCLUSION

Since its inception *Mathematics for Social Justice* has been a welcome addition to the mathematics course offerings at Saint Michael's College, and has become a popular choice among students who need to fulfill their quantitative reasoning requirement but do not need a statistics or calculus class for their major. The social justice theme of the course entices students to enjoy mathematics through investigating problems of genuine relevance. These students are often very interested in the social justice topics of the course and want to find the answers to the associated questions. Furthermore, the modular nature of the course has been helpful for students who have historically struggled with mathematics including those with learning differences that impact their learning of mathematics, since a student who struggles with one unit knows that shortly the class will move on to a different social justice topic that will likely require a different set of mathematical skills. Students also recognize the benefit of becoming a proficient user of Microsoft Excel. Students articulated these positive attributes of the course in course evaluations and reflections.

In order to explore social justice topics thoroughly, the students *need* to do the mathematics. Having an immediate, important, and engaging purpose for doing the mathematics resulted in students *wanting* to do the mathematics, which is the objective for which we all strive. The topic of income inequality, one that is widely reported on by the media, appealed to them as humanitarians, and is mathematically accessible.

ACKNOWLEDGEMENTS

The authors wish to thank Bonnie Schulman from whom some seed ideas came for this course and Brandon Tries from whom some original material was adapted. We are also grateful to Joanna Ellis-Monaghan for her encouragement and valuable input in the preparation of this paper.

APPENDIX

MA 110: Mathematics for Social Justice

Name: ______________________

Estimating the Gini Coefficient

The Gini coefficient measures income inequality in a society. In this activity, we will approximate the Gini coefficient for the United States, Indonesia, and Brazil.

Terminology:

The **Gini coefficient** is computed as the area between the **Lorenz curve** and the **line of equality** (the region A in the picture), divided by

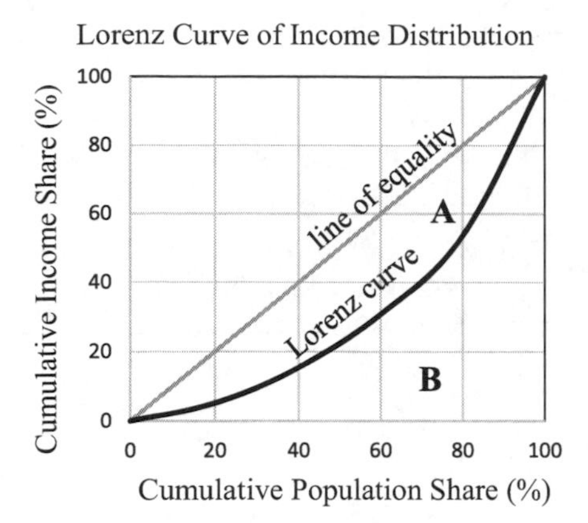

the total area under the line of equality (the region A and the region B, combined). In other words,

$$\text{Gini coefficient} = \frac{A}{A+B}.$$

The **Lorenz curve** gives the proportion of total income held by the bottom $x\%$ of the population (for example, the bottom 10%, might make 5% of the income).

The **line of equality** is the Lorenz curve for a perfectly equal society. (So, the bottom 10% make 10% of the income, etc.) When a society has perfect equality, the Gini coefficient is 0, for perfect inequality the Gini coefficient is 1.

Activity:

1. First, we need data on the proportion of income held by certain percentages of the population. Here is the 2013 data for the United States from the World Bank. To work with the data, we need to make sure that we use the **decimal** rather than the **percentage**.

Percentage of population	Percentage of income (%)	Percentage of income as decimal
Bottom 0%	0	
Bottom 20%	5.1	
Bottom 40%	15.4	
Bottom 60%	30.8	
Bottom 80%	53.5	
Bottom 100%	100	

2. The data in the table above gives a sample of points on a Lorenz curve. We can approximate the area under the curve using trapezoids that go up to these points, as depicted below.
   If we have 5 trapezoids in total of equal width, the width of each trapezoid should be ___________.

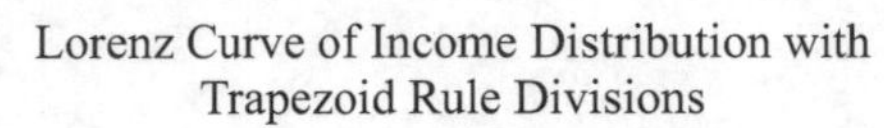

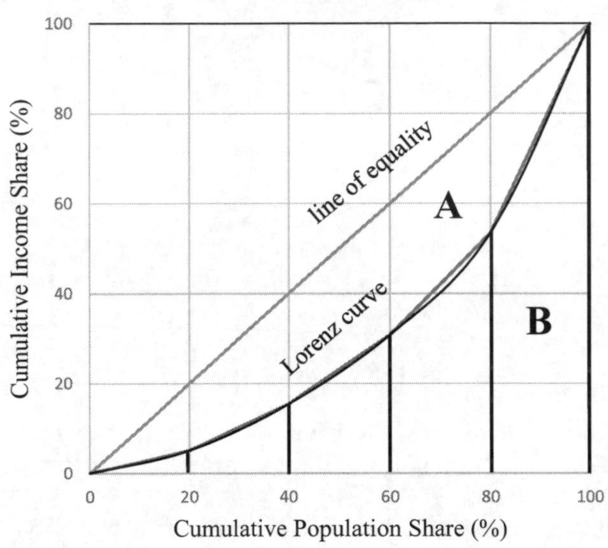

3. The area of a trapezoid can be found with the formula: $A = \frac{1}{2}b(h_1 + h_2)$ where $b =$
   Use the decimal value for the proportion of income for the heights of each trapezoid.

Equation for trapezoid area:	Trapezoid area:
Total:	

4. Now, add up the areas computed in 3. This sum is the approximate area of which region in the figure (A or B)? What is this area?

5. We are almost ready to compute the Gini coefficient. We computed the area *under* the Lorenz curve. We now need to find the area of the region *between* the Lorenz curve and the line of equality. How might we do this? (Hint: the *total* area under the line of equality is 0.5.) Use this to find the area between the Lorenz curve and the line of equality.
6. Divide this by the total area under the line of equality, 0.5, to get the Gini coefficient. What is it?
7. Technology is very useful for doing this type of calculation repeatedly. We're going to check our results by building a table in Excel for the U.S. data.
8. Now, estimate the Gini coefficients for Brazil and Indonesia using the 2013 data from the World Bank given below. (Hint: you may just copy and paste your spreadsheet for the U.S. data onto a new sheet, and replace the data in column B with the data in one of the columns below.)

Percentage of population	Percentage of income: Brazil	Percentage of income: Indonesia
Bottom 0%	0	0
Bottom 20%	3.3	7.2
Bottom 40%	10.9	17.6
Bottom 60%	23.3	31.9
Bottom 80%	42.6	52.6
Bottom 100%	100	100

9. Convert the Gini coefficients to Gini indices (by multiplying by 100). Fill in the following chart. How did we do?

Country	Actual Gini Index	Approximate Gini Index
United States	41.1	
Brazil	52.9	
Indonesia	39.5	

(Source: data.worldbank.org, all data is from 2013)

MA 110: Mathematics for Social Justice

Name: ______________________

Investigation: How has income inequality in the U.S. changed through the decades?

Year	Bottom 0%	Bottom 20%	Bottom 40%	Bottom 60%	Bottom 80%	Bottom 100%	Gini Index
1970	0	4.1	14.9	32.3	56.8	100	
1980	0	4.2	14.4	31.2	55.9	100	
1990	0	3.8	13.4	29.3	53.3	100	
2000	0	3.6	12.5	27.3	50.3	100	
2010	0	3.3	11.8	26.4	49.8	100	

Source: U.S. Census Bureau

Find the Gini index for each of the years using the spreadsheet we created in Excel. The Gini index is the Gini coefficient written as a percent.

Make one meaningful graph that overlays the Lorenz curves for each of the years. Your graph should have a useful & informative title, labels on the axes, and a legend to indicate correspondence. Make sure that I see your graph so you get credit for doing it.

What is the trend of your results? What conclusions can you draw? What are the implications of your results? (Reference the Gini coefficients, the Lorenz curves, and cite from other assignments in your response.)

MA 150: Calculus I

PROJECT 10: INCOME DISTRIBUTION[1]

Names: ______________________

Objectives: Investigate income distribution using integration. Use Maple to fit a power function to data.

How is it possible to measure the distribution of income among the inhabitants of a given country? A powerful tool for illustrating how wealth is distributed across a society is the **Lorenz curve**.

A typical Lorenz curve is given by $y = L(x)$, where $0 \leq x \leq 1$ and $0 \leq y \leq 1$. The variable x represents the cumulative population in the society and $y = L(x)$ represents the cumulative income that is owned by the percent x of the society. For the Lorenz curve shown below, $L(0.5) \approx 0.25$ means that 0.5 (50%) of the society owns 0.25 (25%) of the wealth.

[1]This project was created by merging and adapting Applied Project: The Gini Index from [Stewart, J. (2016). *Calculus: Early Transcendentals*. Cengage Learning.] and Guided Project 19: Distribution of Wealth from [Briggs, Bill; Cochran, Lyle; Gillett, Bernard. (2011). *Calculus: Early Transcendentals*. Pearson.]

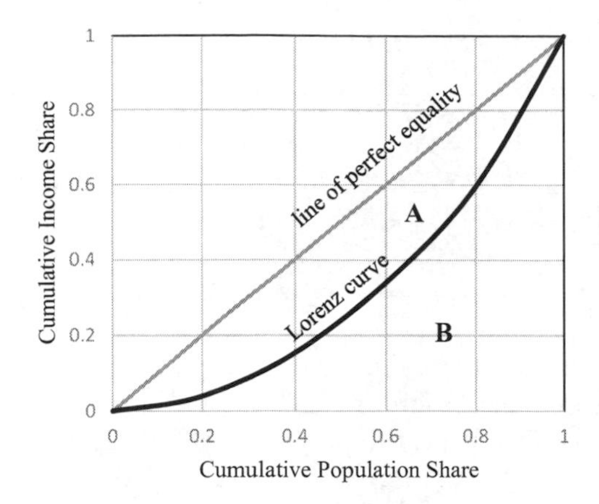

1. In the Lorenz curve shown, $L(0.8) \approx 0.6$. Interpret the point $(0.8, 0.6)$ on the Lorenz curve.
2. A Lorenz curve is always accompanied by the line $y = x$, called the **line of equality**. Explain why this line is given this name.
3. Explain why any Lorenz curve must have the following properties:
   a. $L(0) = 0$ and $L(1) = 1$
   b. L is an increasing function on [0, 1]
   c. The graph of L must lie below or be equal to the line of equality.
4. Power functions of the form $L(x) = x^p$ nicely describe the theoretical Lorenz curve, where $p \geq 1$. Verify that these functions have the properties listed in 3).
5. The information in the Lorenz curve is often summarized in a single measure called the Gini coefficient. The Gini coefficient is the area between the Lorenz curve and the line of equality ($y = x$), labeled A above, divided by the area under the line $y = x$.
   a. Show that the Gini coefficient is twice the area between the Lorenz curve and the line $y = x$, that is

   $$G = 2\int_0^1 [x - L(x)]dx$$

   b. What is the value of G for a perfectly egalitarian society (everybody has the same income)? What is the value of G for a perfectly totalitarian society (a single person receives all the income)?
   The following table, derived from data supplied by the U.S. Census Bureau, shows the value of the Lorenz function for income distribution in the United States from 1970–2010.

x	0.0	0.2	0.4	0.6	0.8	1.0
1970	0	0.041	0.149	0.323	0.568	1
1980	0	0.042	0.144	0.312	0.559	1
1990	0	0.038	0.134	0.293	0.530	1
2000	0	0.036	0.125	0.273	0.503	1
2010	0	0.033	0.118	0.264	0.498	1

6. What percentage of the total U.S. income was received by the richest 20% of the population in 2000?
7. Use Maple to fit a power function to each decade of income data in the table. Then use the power function and the formula from 5) to estimate the Gini coefficient, G, for each decade. Round your final answer to four decimal places.

	$y = ax^p$		
Year	a	p	Gini coefficient
1970			
1980			
1990			
2000			
2010			

1. Create a plot of all the Lorenz curves, along with the line of equality using Maple.
2. Describe the trend of income distribution in the U.S. referencing both the Gini coefficients and the Lorenz curves. What conclusions can you draw? What are the implications of your results?

REFERENCES

1. Briggs, B., L. Cochran, and B. Gillett. 2011. *Calculus: Early Transcendentals*. Boston, MA: Pearson.
2. De Vogli, R., R. Mistry, R. Gnesotto, and G. A. Cornia. 2005. Has the relation between income inequality and life expectancy disappeared? Evidence from Italy and top industrialised countries. *Journal of Epidemiology & Community Health*. 59(2): 158–162.

3. Kawachi, I. and B. P. Kennedy. 1997. The relationship of income inequality to mortality: Does the choice of indicator matter? *Social Science & Medicine*. 45(7): 1121–1127.
4. Lippman, D. 2013. Math in society. http://www.opentextbookstore.com/mathinsociety/. Accessed 15 April 2017.
5. Lopez, R. 2004. Income inequality and self-rated health in US metropolitan areas: A multi-level analysis. *Social Science & Medicine*. 59(12): 2409–2419.
6. OECD. http://www.oecd.org/. Accessed 15 April 2017.
7. Stewart, J. 2016. *Calculus: Early Transcendentals*. Boston, MA: Cengage Learning.
8. U. S. Government. https://www.census.gov/data/tables/time-series/demo/income-poverty/historical-income-inequality.html. Accessed 15 April 2017.
9. Wikipedia. 2017. Gini coefficient. https://en.wikipedia.org/wiki/Gini_coefficient. Accessed 10 April 2017.
10. Wilkinson, R. 2011. TEDTalks: Richard Wilkinson – How economic inequality harms societies. [Video File]. https://www.ted.com/talks/richard_wilkinson?language=en. Accessed 14 April 2017.
11. World Bank. http://www.worldbank.org/?cid=ECR_GA_HPlaunch_searchad_EN_EXTP&gclid=Cj0KEQjw5sHHBRDg5IK6k938j_IBEiQARZBJWnM0s9BII7u02JBXTOk1hT3raH1rOvsJH_9U8ersncMaApht8P8HAQ. Accessed 15 April 2017.
12. YouTube. 2013. Wealth inequality in America: Perception vs reality. [Video File]. https://www.youtube.com/watch?v=vttbhl_kDoo. Accessed 14 April 2017.

Further information

De Maio, F. G. 2007. Income inequality measures. *Journal of Epidemiology & Community Health*. 61(10): 849–852.

Haughton, J. and S. R. Khandker. 2009. *Handbook on Poverty and Inequality (English)*. Washington, DC: World Bank.

Inequality.org. 2012. U.S. income distribution: Just how unequal? http://inequality.org/unequal-americas-income-distribution/.

St Louis Fed. 2010. U.S. income inequality: It's not so bad. https://www.stlouisfed.org/publications/inside-the-vault/spring-2010/us-income-inequality-its-not-so-bad.

“There Are Different Ways You Can Be Good at Math”: Quantitative Literacy, Mathematical Modeling, and Reading the World

K. Simic-Muller

Abstract: This manuscript describes a quantitative literacy course focusing on issues of economic and racial justice, developed for a summer bridge program. The curriculum for the course, described in the manuscript, consisted of open-ended assignments that dealt with real-world issues and required basic modeling skills; whereas the culminating assignment was a final project about a community-based topic. The course helped students deepen their understanding of the issues addressed and strengthen their mathematical reasoning skills; and had a transformative effect on some students’ perceptions of the usefulness of mathematics, of themselves as mathematics learners, and of their communities.

1. INTRODUCTION

In his final address as the President of the Mathematical Association of America, “Mathematics for Human Flourishing” [20], Francis Su reminded the mathematics community of what it means to teach mathematics in our time: in addition to promoting its truth and beauty in our teaching, we must also incorporate play, love, and justice. Su writes, “Justice means: setting things right. And justice is a powerful motivator to action.” There are many paths to making our mathematics community more just, and Su describes a few of them. This manuscript discusses the possibility of creating a just space in the intersection of quantitative literacy, social justice, and mathematical modeling; a space that increases access to success in college mathematics. I will describe a quantitative literacy course that focused on issues of economic and racial justice and that made heavy use of basic mathematical models to make sense of the topics it addressed. In addition to helping students broaden and deepen their mathematical knowledge, the course goals were closely aligned to the traits Su describes: to make mathematics engaging and exciting, and more like play; for students to discover affinity, if not love for mathematics; for the mathematics to serve as a tool for seeking justice; and, through play, love, and justice, for students to begin to see themselves as successful mathematics learners.

2. QUANTITATIVE LITERACY, READING THE WORLD, AND MATHEMATICAL MODELING

Quantitative literacy, also known as numeracy, has been described as "the new literacy of our age" by one of its pioneers Lynn Steen [19]. It has many interpretations, but can be thought of simply as the ability to see the world through a mathematical lens. The opposite of numeracy, innumeracy [15], includes, among other things being: fooled by faulty mathematical or statistical arguments, unable to understand the mathematics needed to make financial decisions, afraid to utilize mathematics in one's life, or convinced that mathematics has no bearing on it. One of the goals of quantitative literacy courses is to help students become numerate citizens.

Very close to the idea of numeracy is the notion of *reading the world with mathematics*, first developed by the Brazilian educator Paolo Freire [8] in the realm of literacy, but then expanded to mathematics by Frankenstein [7] and Gutstein [10]. Reading the world entails understanding the socio-political and historical circumstances of our lives [10], and mathematics provides a language for deepening that understanding. Just as with numeracy, we view the world through a mathematical lens, but we zoom in to systemic injustice, especially as it relates to race and wealth. As the world grapples with the effects of increasing wealth inequality, and as race takes center stage in the national discourse, these topics need to be investigated from multiple angles, including a quantitative one. The literature on teaching mathematics through social justice contexts includes a growing body of curriculum materials at all levels (e.g., [6, 11, 12]), addressing a vast variety of topics such as racial profiling, wealth distribution, or school overcrowding [11].

Reading the world with mathematics does not come with a neat set of rules, and one should not expect questions that arise in interrogating the fairness of the world to be neat and answerable using predetermined algorithms. In fact, they often involve mathematical modeling. Anhalt [1] identifies the following traits of modeling problems: they are open-ended, use contexts that are less contrived, no algorithm is associated with them, not all parameters are given, little structure is provided, assumptions need to be made, and approximate solutions are acceptable. Students who have not been successful in traditional mathematics classes are often successful in modeling, as it requires different skills and does not need a strong algorithmic background [13].

2.1. Affordances and Challenges of Reading the World With Mathematics

Pedagogy built around reading the world with mathematics needs to be implemented carefully, and has been used with varying degrees of success. A number of educators report high student engagement, success in solving the open-ended problems posed, and increased awareness of issues (e.g., [6, 7, 10]). However, it is important to acknowledge the challenges inherent in this work: creating curriculum built on real-world data is time-consuming [4]; some topics cannot be easily connected to real-world issues [4]; there may not be enough time to address these contexts in depth [9]; or, conversely, conversations may focus too much on the real-world issue and not on the mathematics itself [4, 5]. Finally, students may be resistant to addressing real-world contexts, as they believe that mathematics should be kept separate from societal issues [4], because they are tired of representing their entire race or culture to the dominant group [18], or because they are the dominant group and refuse to grapple with their complicity in the system [17]. In other words, although the pay-off may be high, so is the investment. Instructors need to be versed in the topics they are discussing; in the strengths and struggles of their students of their communities, and in facilitating discussions: all skills for which mathematicians typically do not receive training. They have to constantly reflect on their successes and failures, and expect to grow with the students [10].

3. COURSE DESCRIPTION

The course described here was part of a 5-week residential program at a medium-sized private liberal arts university in the Pacific Northwest. It was part of a summer bridge program for incoming first-year students who were identified as needing additional academic support. Students were placed in my class either because they had low mathematics placement scores or because their majors did not require any more mathematics. These are students typically dismissed as "weak" and placed into remedial courses. The class described here was in no way remedial and built on students' strengths while systematically addressing gaps in previous knowledge. In order to improve the students' relationship to mathematics, it was essential in this class never to approach their backgrounds as being deficient, but to mediate the gaps in their knowledge in multiple ways, especially through teamwork, the use of technology, and carefully scaffolded instruction.

We met four days a week for 3.5 hours. In addition to mathematics content, class time was spent doing team-building activities, discussing strategies for succeeding in mathematics classes and in college, and planning activities students later implemented in the neighboring middle school's summer program. These activities also helped build trust between the students and me, which is an essential prerequisite for what Gutstein calls "co-creating a classroom supporting social justice" [10].

The mathematical topics covered were standard for a quantitative literacy or liberal arts mathematics course, and included ratios, rates, and conversions; percentages; basic statistics; and linear and exponential functions. Excel was an essential part of most classroom investigations, and it was especially useful in allowing students to engage with basic mathematical modeling. Excel is particularly beneficial in allowing students who may struggle with the computational aspect of algebra to engage in higher-level mathematical thinking. We used *Common Sense Mathematics* [3] as text, and I supplemented the textbook with my own lessons and activities that were relevant to our setting. I largely used curriculum my colleagues and I had developed [6], along with some activities from [11] and [14]. I also created some new lessons to accompany activities that were taking place elsewhere in the program. A large component of the course was the final project, which required students to pick a topic that would be personally relevant to them and also relate to the community, for example, the campus, school district, neighborhood, or city.

Although it was helpful to have a small class and additional support that a bridge program provides, the course described here could easily be adapted to fit the format of a regular quantitative literacy or liberal arts mathematics course. Furthermore, it is possible to adapt ideas presented here to other mathematical content areas.

4. CURRICULUM

Social justice contexts were central to the class, but we did not engage with them all the time. I also created space in the classroom for mathematical play, to heal some mathematical wounds many of the students carried: we played games and engaged in hands-on activities. Equally important was care. Without unconditional acceptance for the students and their mathematical knowledge as it was and not as I wished it to be; and without an understanding that the students were experts in their lives and of the effect that the issues we discussed had on them, this class would not have been possible.

As these were incoming first-year students, we discussed topics such as reading a mathematics textbook, studying for an exam, or writing in mathematics. We also started homework in class so students would be less likely to give up when faced with an unfamiliar problem. Finally, even though the focus of the course was unequivocally on open-ended real-world problems, based on a diagnostic test that students completed on the first day, the class worked on review worksheets for every mathematical topic discussed in class. These worksheets provided refreshers in mathematical procedures needed for real-world problem solving.

4.1. Preparing for Modeling

As mathematical modeling is typically foreign to students, who have little prior experience solving ill-defined, open-ended problems, our introduction to modeling was gradual. Initially we engaged with problems that had only a subset of traits of a modeling problem.

On the third day of class, students participated in a role-playing simulation named *Archie Bunker's Neighborhood* [2]. In the simulation, each participant was assigned a different race, religion, or ethnicity, and was treated according to stereotypes about that group. Consequently, many experienced difficulties in the simulation, including in obtaining housing. The simulation was emotional, but not obviously based on facts. Since problem posing is important for modeling, the homework for the following day asked students to pose their own mathematical questions related to the simulation, and they came up with some relevant ones, including: *What is the difference in percentages between White men and Black men imprisoned in the U.S.? What is the median household income with in each of the communities we looked at? At what rate do immigrants move into communities? At what rate do communities grow in population?* At this time, students only posed questions and did not attempt to answer any of them. We would investigate many of these questions in some form in the following weeks.

On the next day, we investigated mortgage denial rates, by race, in different U.S. cities. In particular, we looked at the disparity ratio between denial rates for Black and White mortgage applicants. We looked at the table found at [22] and discussed an open-ended question,

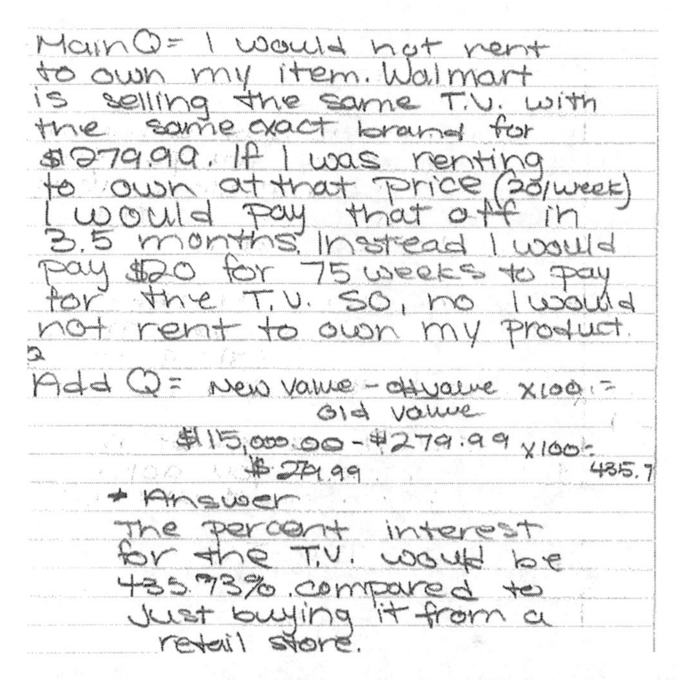

MainQ= I would not rent
to own my item. Walmart
is selling the same T.V. with
the same exact brand for
$279.99. If I was renting
to own at that price (20/week)
I would pay that off in
3.5 months. Instead I would
pay $20 for 75 weeks to pay
for the T.V. So, no I would
not rent to own my product.
2
Add Q= New value - old value x100 =
old value
$15,000.00 - $279.99 x100=
$279.99
435.7
* Answer
The percent interest
for the T.V. would be
435.73% compared to
just buying it from a
retail store.

Figure 1. Rudy's and Lili's calculation and conclusion for a Samsung TV.

What are some observations you can make about the numbers in the table? a more mathematically focused one, *The ratio obtained when you divide the denial rate for Black applicants by the denial rate for White applications is called a disparity ratio. What do you think that number tells you? Which cities have the highest disparity ratios?*, and a question that would make sense of the data: *What might be some reasons for higher denial rates for non-White applicants?* We discussed the questions as a class, but this would become the typical format of most of the lessons: an open-ended question at the beginning followed by some mathematical investigations and a reflection that summarizes the mathematics and allows for personal stories to be shared at the end.

Next, we investigated a task that provides some foundation for modeling but provides more scaffolding than that typically encountered in modeling a problem. We investigated under what circumstances a minimum wage earner could afford an apartment in different cities in Washington State, Oregon, and Alaska, the three states from which all the students in the class originated. Information provided included minimum wage and rental information, and the federal guidelines on affordability of housing [16], which state that rent should be no more than 30% of gross income. In all but one county investigated, one minimum wage

Reflection Q =
This investegation tells
me about the rent to own
buisness is that you don't
always get the best deals off
rent to own. You utimatelly
pay more $$ renting to own
because you pay interest on
the product. This beneficial
for me because I know now
that it is a lot cheaper to
"buy now" then make payments
+ interest. People use rent to
own buinisses to build
up their credit score. To show
that their credible of paying
off something. Also if they
don't have the money upfront,
they can at least pay a %
of it there.

Figure 2. Rudy's and Lili's conclusion about rent-to-own businesses.

earner could not afford a one- or two-bedroom apartment at the average price in that area. Because all students had some prior knowledge of and opinions about the topic, and because all investigated their hometown, the engagement was high. Prior experiences often added another dimension to the activities, which were necessarily more simplistic than real life. Rudy (all names are pseudonyms), grew up in Section 8 housing, where rent is based on income, and was therefore confused as to why we were determining affordability of an apartment based on whether it cost less than 30% of a minimum wage earner's gross income. In his calculations, he found 30% of the rent, noted that this was how much the family would be paying, and decided that the apartment was affordable in all cases. Although this response did not match the one I expected, it provided a learning experience for me in assumptions that are made in creating curriculum.

The first completely open-ended assignment was one where students investigated whether to rent to own an item based on a flyer from a rent-to-own business, and using the percent change formula and calculating interest to determine their answer. Figures 1 and 2 show Rudy's and Lili's work on this assignment. Rudy's work is featured throughout, not because it was always correct or most advanced (it was not), but because Rudy experienced a transformation in the course, and because his personal experiences had an impact on the curriculum, as in the example above.

4.2. Linear Functions

To engage properly with basic mathematical modeling, students need to make predictions in addition to answering questions. Linear functions are a good tool for this, yet provide some remarkably poor predictions when used long-term. To make sure that students would know how to write equations as well as make predictions, we did two investigations with linear functions: one about the cost of water in Flint, and the other about the change in incarceration rates since 1925.

4.2.1. Cost of Water in Flint

I developed this activity from a news story I read shortly before I taught this class [21]. In the activity [6], we used linear functions to compare water bills in Flint and our own community (see Appendix A for the prompts). I presented the class with my water bill and students wrote an equation for my water bill as a function of water used. Based on this equation, they looked at how much it would cost to use 60,000 gallons of water in a year, and how much water my family would use to have an annual bill of $864. These figures were not chosen randomly: in 2015, at the height of the water crisis, a household in Flint would pay $864 for 60,000 gallons of water, the highest cost in the country [21]. The activity was mostly guided, but included decision making about whether to write equations based on monthly or annual use and how to convert between cubic feet and gallons. It also provided an authentic real-world application of the slope-intercept form of a linear function, as a water bill includes fixed and variable costs. The students, who had known a little about the Flint water crisis, were shocked by the results of the activity. Even though they struggled with writing the equations, they were so genuinely surprised by the outcome of the activity that they found the struggle worthwhile.

4.2.2. Incarceration Rates

In this two-part activity, created by a colleague [6], students were given a spreadsheet with incarceration data from 1925 to 1970. The only prompt that accompanied the data was *Predict what the prison population would be in 1980, 1990, 2000, 2010, and 2020.* Students created linear models to predict the current prison population based on these data, but because the prison population in the U.S. exploded starting in the 1970s, their predictions were vast underestimates of the true figure. This gave us an opportunity to discuss limitations of linear models. In the second part of the activity, students were given another spreadsheet, this time with incarceration data up to 2006, and modified their predictions using an exponential model. The last prompt was *What have you learned from this investigation? What else do you want to know?* This gave the class the opportunity to reflect on the trends and pose additional questions about incarceration in the U.S.; this was one of the topics about which many were curious after the Archie Bunker's Neighborhood activity. They also compared total numbers and percentages of incarcerated population by race [14].

Students sometimes struggled with the mathematics or with interpreting their results. For example, they were unsure why the numbers of incarcerated Black and White men were almost equal even though they knew to expect incarceration to have a larger impact on Black communities. Even when struggling with calculations, they were able to note that the incarceration rates were disproportionate, displaying critical thinking. For example, Antony incorrectly calculated the percentages of each segment of the population incarcerated, but correctly concluded the following:

> Black people are disproportionately being put into jail when comparing the rate of incarcerated people to their population of race. White people make up 5 times the amount of black people in terms of the general population, yet their incarceration rates are significantly lower when contrasting the black population.

Similar observations abounded throughout the course, showing developing numeracy, and increasing the critical awareness of issues with which we were dealing. Other modeling assignments included predicting future U.S. and world population levels and global temperature change [3], and investigating immigration, unemployment, and crime trends.

4.3. Final Project

The culmination of the course was the final project. Students picked a topic relevant to them, and preferably to the university and its surroundings, and used the mathematics utilized in the course to investigate this topic. Everything else that the students did in the program (e.g., learn about the surrounding communities, unpack injustice related to socioeconomic status and race) or thought about (e.g., "How does this community compare to mine?" or "What is my place here?") would ideally come together in the project. They were not required to pick a social justice-related topic, but all did. Topics investigated included: the gender income gap; sexual assault on college campuses; comparison of two school districts in terms of enrollment, poverty, and free and reduced lunch eligibility; gentrification; and crime rates. The projects allowed the students to put their developing mathematical modeling skills into action and to learn more about a topic of personal interest.

As writing papers in a mathematics class is uncommon, I provided many checkpoints for the final project. Students submitted increasingly detailed documents every week: first just a proposed topic with some mathematical questions, then one-page proposal, outline, first and second draft, and eventually the final draft of the paper. I gave constructive feedback at every step of the process. Only the final draft was graded according to the rubric (presented in Appendix B with detailed instructions for the project).

Rudy, who grew up in the inner city and perceived it as unsafe, chose crime in the city as his topic, and his initial questions were: *What percentage of the homicides are by minorities? What percentage of the homicides are by convicted felons? How does the number of drug crimes affect the number of homicide rates? How does the number of minorities under the poverty line effect [sic] homicide rate?*

In response, I wrote to Rudy:

> These are really interesting questions but I wonder how you will find the data to answer them. You may find that you change your questions as you go. You may have to look at national data if local data is hard to come by.

I also sent him some websites to use in his research.

The following week, using the website I shared with him, Rudy reported on the total number of homicides and the homicide rate for the city, state, and entire United States between 2003 and 2014. He ended the proposal with the following quote:

> I learned through this website that the crime rate in total has decreased tremendously since the late 20th century. I can use this website to answer my question by using more in depth statistics ... to be more critical in my final project.

The week after, Rudy added more tables. Then, for the final draft, he included calculations, graphs he created, and interpretation of his results. In the process, his questions changed, due in part to a lack of access to certain data, but also due in part to a changed perception of crime. The questions eventually became: *Is my hometown the most dangerous city in the state? What city in the state has the highest Violent Crime Rate? How does the city's violent crime rate compare to the National Average and state?*

In the final paper, Rudy made sense of how violent crime rates are calculated and then looked at trends in homicides, rapes, robberies and assaults over the past 12 years. Whereas other students might have used more intricate mathematics in their reports, Rudy's report was more personal, and even passionate. He was initially skeptical of findings that crime has been decreasing for the past 30 years. However, in the conclusion to his paper he wrote:

> Even though [the city] has a very high violent crime rate, its numbers are decreasing. In every form. Homicides, rapes, robberies and assaults all have been decreasing over the past 10 years. [The city] may not be the safest place to live but it is getting safer, which is all we can ask for.... But is still way to [sic] high, to stop letting up the pressure to stop these crimes from occurring.

4.4. Assessing Student Work

As real-world investigations primarily took place during class, and were done in groups, students had ample opportunities to modify their solutions and models as needed. Typically, by the end of class, with some assistance, groups would come to a correct mathematical solution. I also gave students opportunities to redo incorrect assignments. This made assessment time-consuming, but mostly straightforward. I did not mark down reflections if they did not have the conclusion that I wanted, and would accept any answer that was based on correct mathematics and was a product of careful consideration of the facts. Consider the following two reflections on the assignment where students discovered percentages of incarcerated White, Black and Hispanic/Latino Americans.

> Reflection 1:
>
> The percentages of black, white, and hispanic people in prison are proportionally unequal. Although white people make the highest percent of the general population, they make the lowest percent when it comes to incarceration. Meanwhile, black people, who only make up approximately 12% of the population make up nearly half of the US prison population. This reveals a HUGE inequality between incarceration among different races.
>
> Reflection 2:
>
> This worksheet tells me the statistics of how many white/black/hispanics [sic] are in prison – however there are more people in this world then [sic] those three races. This worksheet also doesn't tell me why people are in prison. Are they all serious offenders? Some misdemoners [sic]? How many are men and how many are women?

The first reflection is exemplary, as it correctly identifies the main point of the activity. The second reflection should not be dismissed either, even though it misses this point, as it offers opportunities for continued conversation and raises additional questions rather than just agreeing with the conclusion. However, had a reflection completely ignored the disproportionate incarceration of people of color, I would have followed up with additional questions to bring the student's attention back to the issue.

On all formative assessments, multiple drafts and opportunities for corrections were crucial. Similarly, reflections offered an opportunity for continued dialog, which is why giving detailed and timely feedback on all assignments was essential in this class.

The quizzes and tests were more traditional, though they investigated the same topics just in a more structured manner, and students were allowed to use Excel for any of the problems. In particular, one of the questions on the final exam returned to mortgage denial rates and asked students to create histograms to compare denial rates for Black and White applicants.

5. DISCUSSION

5.1. Reflections

As in Gutstein [10], I employ reflections as a form of assessment. Both at the beginning and at the end of the course, students responded to the following five questions, in essay form:

1. What do you think about yourself as a mathematics learner?
2. What does it feel like to do mathematics?
3. How do you best learn mathematics?
4. Do you think math is relevant to your life? How?
5. Do you think math is relevant to your future career? How?

Additionally, at the end of the course, students were asked to look at their original reflections and write about any changes since the beginning of the class. All agreed that mathematics was much more useful than they had previously thought, and all (albeit grudgingly) admitted that they were better at math than they had thought. Rudy wrote:

> I have learned mostly how to think independently, and to use more critical thinking. It feels good to do mathematics, in this class I learned that mathematics is so much more than you expect.

Kristin wrote:

> Now that I have taken this math class I think differently of myself as a math learner. There are different ways that you can be good at math,

> just because I didn't take Calculus in high school and I did average in stats doesn't mean I am terrible at math, it just mean that I have a different set of skills than those who took calculus .... I think I best learn math when connecting it to real life concepts and social justice issues like we have been doing in class. It connected to my major too which made me find the class more interesting, relatable, and useful. Math is relevant to my life and my major, I learned to not trust all statistics I see and to do the math and research myself and then make my decision on whether or not to trust it.

Kristin's reflection mirrors the goals for the course stated at the beginning: increased appreciation of mathematics, increased confidence in herself as a learner, and awareness of mathematics as a tool for reading the world.

Ultimately, all students were successful in the class. We could measure their success in grades: all passed it with a grade of B- or higher, and did moderately well on the tests and quizzes. But certainly, these grades could be inflated and besides success should not be measured in terms of scores. I deem the students successful because they learned to create solid mathematical arguments and because they were willing to engage with difficult topics and messy mathematics that stretched their perceptions of what mathematics can be. There is no reason to assume that students with mathematical backgrounds that are labeled as weak are unable to engage with higher-level investigations. In fact, because mathematical modeling often requires critical thinking as much as, if not more than, calculations; and as tools such as Excel can help with calculations, all students can be successful in a quantitative literacy course with sufficient scaffolding and review.

5.2. Challenges

The goal of this manuscript was to demonstrate what students are capable of in a just space created in the intersection of quantitative literacy, social justice, and mathematical modeling. However, it is important to acknowledge challenges in creating such a space. First, as documented in research [4], designing this course was time-consuming, and required daily adjustments based on what was happening in and out of the classroom. Although the effort was definitely worth it, instructors need additional time for finding appropriate data sets or relevant readings, and for providing timely and detailed feedback on all assignments. Second, as expected ([4, 17]), there was some resistance to contexts we discussed. Most students in this class belonged to one or more marginalized groups, and their lived experiences were often mirrored in the curriculum, but not to the extent that no new learning took place. For the few with mostly dominant identities, it was not always easy to engage with the contexts. Resistance was subtle, and mostly visible in lack of motivation and deficit perspectives expressed in reflections. In my experience, it is essential to provide space for reflection and dialog, and to always let mathematics be the authority. Although the causes of high incarceration rates can be debated, its disproportionate impact on communities of color cannot. Finally, five weeks are simply not enough to gain a deep understanding of any topic, whether it be mathematics content or a real-world context. Regardless of the amount of scaffolding, not all students produced excellent final papers. A few did not contain much mathematics, and it is always challenging for students to create mathematical arguments based on data and computations. At least one student left the course with a deficit view of the surrounding community. Nevertheless, the course planted seeds in students' minds that could, in a proper environment, blossom in the future.

5.3. Advice to Instructors

The first of three practices that Gutstein [10] identifies as essential in co-creating social justice classrooms is normalizing politically taboo topics. If we are tentative about addressing these topics, students will resist them. If we present them as genuine mathematical problems to solve and debate, rather than agree on, students are generally open and curious. This can be challenging, and I will note that I have been practicing having brave conversations about difficult topics for years. I have found that the only way to become a competent facilitator is through constant practice, increasing knowledge of topics we are discussing, and openness to listening to and learning from students.

In planning a course such as this one, it is important to find a balance between topics that we know and are personally invested in, and ones about which students care. Although I made many curricular decisions based on my overarching goal for the course, I also included topics based on student interest. When at the beginning of the semester incarceration came up in conversation, I included two activities about incarceration rates, which I may not have done if the interest had not been there.

Quantitative literacy courses and liberal arts mathematics courses often provide a flexible structure that allows for incorporating justice-oriented contexts. However, justice has a place in all mathematics classrooms, and mathematical modeling investigations similar to ones described here can be modified for other courses, most notably College Algebra, Precalculus, and Calculus. Other instructors have been successful in teaching such courses [12].

6. CONCLUSION

If mathematics is to be a tool for human flourishing, it has to be rooted in justice. We can, and should, seek justice through mathematics. For example, knowing that water in Flint, while being tainted with lead, was also twice as expensive as it is where my students live is a step toward justice. Even if we do not act on it, the fact that our eyes are open to this fact, the fact that we are reading the world, is the beginning of seeking justice. However, justice also means redistribution of power. Justice is allowing students' voices to be heard and their stories to be told. This is not only the purview of humanities: our mathematical stories can be just as impactful. For example, having grown up in a troubled city, Rudy is wary of it. He does not trust it the way that I do because we did not live the same experience. His stories are valuable. The mathematical story of declining crime is also powerful, and allowed him to embrace said city in the end, as flawed but loved and changing. This is possible because of a space we created in the classroom, one where students were mathematicians solving real mathematics problems that were relevant to them and to the community they would soon call home. In this space, in order to hear student voices clearly, we must always focus on what they know rather than on what they cannot yet do.

APPENDIX A: COST OF WATER IN FLINT

On my monthly water bill, I pay \$22.62 for fixed monthly costs, and pay \$1.756 per CCF used. (Note: 1 CCF = 748 gallons.)

1. Why are there fixed monthly costs on my water bill? What do they represent?
2. Write an equation for the amount of dollars (y) that I will pay for water per year if I use x gallons of water. Note that you will have to convert from CCF to gallons.
3. Use Excel to answer the following question: How much would I be paying for water per year if my family used 60,000 gallons per year? (Note: It is estimated that an average U.S. family uses about this much water per year.)
4. Use Excel to answer the following question: How much water would I have used if my annual water bill is \$864?
5. Do you think that Flint, Michigan residents, on January 1, 2015, had higher or lower water bills than I did? Why do you think that?
6. Make a guess as to how much a household using 60,000 gallons of water was paying for water in Flint, Michigan in 2015.

APPENDIX B: FINAL PROJECT INSTRUCTIONS

What is this project about?

The goal of this project is for you to use mathematics to understand the topic of your choice better or differently than you have in the past.

This project can be about one of the following: the campus, the neighborhood it is in, the city it is in, the neighborhood middle school, the community you come from, and especially about a combination of some of these topics.

How do I pick a topic?

Here are some suggestions for picking your topic:

- The best way to start is to think about something that you have learned in another segment of the Summer Academy. For example, was there anything that intrigued you during the walking tour of the neighborhood? Your guide asked many thought-provoking questions that were related to mathematics. Is there anything that you learned in one of the other two classes you are taking that you want to investigate using math? Do you want to know more about middle school students and their education?
- Is there anything about your community that you want to investigate through math? Is there something that you wish that people knew about your community? Can you compare your community to this one? (For example, graduation rates at your high school, ethnic breakdown of your neighborhood, house prices, etc.)
- Is there anything about the university that you would like to know that you haven't learned yet? (e.g., class sizes, diversity, etc.)

There are many possibilities. Make sure to pick a topic that you actually care about.

What do I do when I pick a topic?

When you identify your topic of interest, ask yourself: How is this topic related to math? Then conduct some research and put together a report and presentation that will show how your topic is related to math.

This may seem difficult, and you may not see connections between math and other areas of your life and learning. However, just about every topic is related to mathematics in some way, if you know what to ask and where to look. We will look at a lot of applications of math to real life in this class, which might also help you with your research. If you are having difficulties identifying the connection, you can either change the topic or come to office hours for help.

Because the project is about your interests, you will be working on your own, though you are welcome to discuss your project with and get advice from your classmates.

How should I conduct research?

Once you have a topic, the easiest (though not necessarily the most accurate) way to conduct research is via the Internet. Be careful which sources you use. Whenever possible, use official sources, for example Census data, The National Center for Statistics, or newspapers like The News Tribune or your local newspaper (though even they are not truthful and accurate all the time). Wikipedia can be considered a valid source of information, but you cannot use it as the primary one. Once you have chosen a topic, I can also guide you to appropriate websites and readings.

It might also be helpful to talk to people. For example, you can talk to people in the community or contact appropriate people on campus to get your questions answered.

Try to stick to your project topic throughout to make your life easier. If you need to switch, please talk to me first.

How do I ask and answer a math question?

Say you want to learn more about graduation rates. You can research graduation rates in different states or in different school districts in Washington, or in different schools in the area and compare them. You can compare the graduation rates for White, Black, Asian, Hispanic, and American Indian students, and create graphs and tables to show your comparison.

- There has to be a mathematical question that motivates the paper, for example, How do graduation rates for different ethnic groups compare in Washington State? or How have populations of different ethnic groups changed in this community over time?

- The question has to be meaningful to you; its answer has to provide real information that you genuinely care to learn. Your question should be fun or moving to answer.
- The question has to be answerable. You may not be able to find out the graduation rates for your high school, or about neighborhood population before 2000. The question needs to be answerable using mathematics you know.
- The question has to be answered using mathematics, for example calculate percent change in graduation rates for each racial/ethnic group or create statistical graphs to show change in population.

I understand that what I am asking you to do is not easy. You have probably never written a math paper before, and the assignment is extremely open-ended. However, because you are turning in four pieces or drafts of the assignment before the final one, we will be able to have a conversation about your project as it evolves, and I will help you pose questions and look for answers to them.

Formatting the paper

- Please email me all your drafts, as you can receive feedback much faster this way.
- The final project needs to be typed and saved as a Word document (.doc or.docx), including formulas, graphs, and tables.
- The paper should be 5-10 pages long.
- The paper should follow the general format of an academic paper, and include an introduction, exposition, and conclusion, as well as bibliography in the end.
- Please include graphs and tables whenever appropriate.
- Your exposition has to be clear, yet contain in-depth information about the topic. Imagine you are writing for a reputable daily newspaper.
- I do not have a preference regarding the formatting style you will use, but I do need you to cite your sources, both in-text and at the end. Rcmember not to quote extensively, but only when absolutely necessary. You will have to use at least three different sources, at least one of which is not a website (though you can use online editions of newspapers, magazines, and academic journals).
- I will grade you on spelling, grammar, and style, as well as on the mathematical content.

Final project grading rubric

	4	3	2	1
Paper layout	Paper uses at least three references correctly cited, writing is clear, paper is free of grammatical and spelling errors, and information presented in logical order	One of the four items incomplete or missing.	Two of the four items incomplete or missing.	Three of the four items incomplete or missing.
Mathematical content	Mathematics is in forefront of paper and clearly helps understand topic; all mathematics in paper is explained.	Mathematics mostly in forefront of paper, most mathematics present in paper is explained.	Mathematics not the main focus of paper and is difficult to follow, and explanations are mostly unclear.	Mathematics hardly present and is poorly explained.
Mathematical correctness	All mathematics in paper is correct and appropriate for level of class.	Most mathematics in paper is correct.	There are many mistakes in paper; mathematical content is too easy or too difficult.	Mathematics in paper completely incorrect, completely trivial, or completely incomprehensible.
Using visuals	Paper contains at least one graph or table used appropriately that helps understand topic.	Paper contains at least one graph or table, but with some mistakes.	Paper contains at least one graph or table, but is not appropriate or is incorrect.	No visuals in paper.
Justifying claims	All claims backed with mathematical evidence, spcific examples, and information from sources used.	Most claims backed with mathematical evidence, specific examples, and information from sources used.	Some claims backed with mathematical evidence, specific examples, or information from sources used.	Almost no claims backed with evidence, specific examples, or information from sources used.

(*Continued*)

(*Continued*).

	4	3	2	1
Applications to real life	Paper clearly describes relevance of topic to real world and to author of paper.	Paper mostly describes relevance of topic to real world and to author of paper.	Paper states but does not describe relevance of topic to world and to author of paper.	Paper does not mention relevance of topic to real world and author of paper.
Understanding of topic	Author of paper demonstrates understanding of topic and makes no errors.	Author of paper demonstrates understanding of topic and makes some errors.	Author of paper demonstrates partial understanding of topic and makes multiple errors.	Author of paper shows no understanding of topic and makes repeated errors.

Timeline

- **Due July 6**: Pick a topic you are interested in and ask some mathematical questions about it. Your classmates and I will give you feedback in class.
- **Due July 12**: Write a proposal for your final project. Include some questions that you want to answer, and conduct some research to try to answer some of your questions.
- **Due July 19:** First draft is due.
- **Due July 26:** Second draft is due.
- **Due July 28:** Final draft of the project and the project presentation are due.

Each part of the assignment will get either partner feedback or whole class feedback in addition to mine. You will also have to meet with me individually to discuss the project.

REFERENCES

1. Anhalt, C. O. 2014. Scaffolding in mathematical modeling for ELLs. In M. Civil and E. Turner (Eds.), *The Common Core State Standards in Mathematics for English Language Learners: Grades K-8*, pp. 111–126. Annapolis, MD: Tesol Press.
2. Archie Bunker's neighborhood. https://sjsummit.wikispaces.com/file/view/Archie | Bunker.pdf. Accessed 31 March 2017.
3. Bolker, E. and M. Mast. 2016. *Common Sense Mathematics*. Washington, DC: Mathematical Association of America.

4. Bratlinger, A. 2013. Between politics and equations: Teaching critical mathematics in a remedial secondary classroom. *American Educational Research Journal*. 50(5): 1050–1080.
5. Felton, M. 2010. Is math mathematically neutral? *Teaching Children Mathematics*. 17(2): 60–63.
6. Felton-Koestler, M., K. Simic-Muller, and J. M. Menéndez. 2017. *Reflecting the World: A Guide to Incorporating Equity in Mathematics Teacher Education*. Charlotte, NC: Information Age Publishing.
7. Frankenstein, M. 1998. Reading the world with math: Goals for a critical mathematical literacy curriculum. In E. Lee, D. Menkart, and M. Okazawa-Rey (Eds), *Beyond Heroes and Holidays: A Practical Guide to K-12 Anti-racist, Multicultural Education and Staff Development*, pp. 306–312. Washington, DC: Network of Educators on the Americas.
8. Freire, P. and D. Macedo. 1987. *Literacy: Reading the Word and the World*. South Hadley, MA: Bergin & Garvey Publishers.
9. Gregson, S. 2013. Negotiating social justice teaching: One full-time teacher's practice viewed from the trenches. *Journal for Research in Mathematics Education*. 44(1): 164–198.
10. Gutstein, E. 2006. *Reading and Writing the World with Mathematics: Toward a Pedagogy for Social Justice*. New York, NY: Routledge.
11. Gutstein, E. and B. Peterson. 2013. *Rethinking Mathematics: Teaching Social Justice by the Numbers*. Milwaukee, WI: Rethinking Schools.
12. Karaali, G. and l. Khadjavi. In press. *Mathematics and Social Justice: Perspectives and Resources for the College Classroom*. Washington, DC: Mathematical Association of America.
13. Lesh, R. and R. Lehrer. 2003. Models and modeling perspectives on the development of students and teachers. *Mathematical Thinking and Learning*. 5(2&3): 109–129.
14. Osler, J. 2007. Radical Math: Statistics Unit, Part 1 (Teachers Version). radicalmath.org.
15. Paulos, J. A. 1988. *Innumeracy*. New York, NY: Hill and Wang.
16. Simic-Muller, K. (in press). Who makes the minimum wage? In G. Karali and L. Khadjavi (Eds), *Mathematics and Social Justice: Perspectives and Resources for the College Classroom*. Washington, DC: Mathematical Association of America.
17. Sleeter, C. 2001. Preparing teachers for culturally diverse schools: Research and the overwhelming presence of whiteness. *Journal of Teacher Education*. 52(2): 94–106.
18. Solorzano, D., M. Ceja, and T. Yosso. 2000. Critical race theory, racial microaggressions, and campus racial climate: The experiences of African American college students. *The Journal of Negro Education*. 69(1/2): 60–73.
19. Steen, L. A. 1997. *Why Numbers Count: Quantitative Literacy for Tomorrow's America*. New York, NY: College Entrance Examination Board.
20. Su, F. 2017. Mathematics for human flourishing. https://mathyawp.wordpress.com/2017/01/08/mathematics-for-human-flourishing/. Accessed 29 March 2017.

21. Wisely, J. 2016. Flint residents paid America's highest water rates. http://www.freep.com/story/news/local/michigan/flint-water-crisis/2016/02/16/study-flint-paid-highest-rate-us-water/80461288/. Accessed 19 January 2018.
22. Zillow. 2015. Mortgage denial rates down, especially among black borrowers http://zillow.mediaroom.com/2015-11-05-Mortgage-Denial-Rates-Down-Especially-Among-Black-Borrowers. Accessed 19 January 2018.

The Brokenness of Broken Windows: An Introductory Statistics Project on Race, Policing, and Criminal Justice

Jared Warner

Abstract: We describe a semester-long project for an introductory statistics class that studies the broken windows theory of policing and the related issues of race, policing, and criminal justice. The most impactful feature of the project is the data-collection phase, in which students attend and observe a public arraignment court session. This "Court Monitoring Project" was completed in partnership with the Police Reform Organizing Project, a New York City non-profit organization. The mathematical student learning outcomes of the project emphasize the construction and interpretation of various graphical representations of data (contingency tables, bar charts, histograms, box plots, and scatter plots).

1. INTRODUCTION

In 1982, criminologist George L. Kelling and political scientist James Q. Wilson introduced the broken windows theory of policing strategy in an article for *The Atlantic* called "Broken Windows: the Police and Neighborhood Safety" [23]. Their theory is that civil disorder and urban decay are harbingers of more serious crimes, and so crime-prevention is best pursued through order-maintenance policing as opposed to crime-fighting policing. They illustrate their argument using the image of the windows in a building: "if a window in a building is broken and left unrepaired, all the rest of the windows will soon be broken." Thus, they argue, police should focus their efforts on repairing the "broken windows" of a community, which they describe as "panhandlers, drunks, addicts, rowdy teenagers, prostitutes, loiterers, [and] the mentally disturbed."

In their article, Kelling and Wilson anticipate a potential shortcoming of broken windows policing that has become one of the most controversial aspects of their theory:

> The concern about equity is more serious. We might agree that certain behavior makes one person more undesirable than another but how do we ensure that age or skin color or national origin or harmless mannerisms will not also become the basis for distinguishing the undesirable from the desirable? How do we ensure, in short, that the police do not become the agents of neighborhood bigotry?
>
> We can offer no wholly satisfactory answer to this important question. We are not confident that there is a satisfactory answer except to hope that by their selection, training, and supervision, the police will be inculcated with a clear sense of the outer limit of their discretionary authority.

Kelling and Wilson were prescient to have this "concern about equity." In [6], Jeffrey Fagan and Garth Davies demonstrate that since the implementation of broken windows policing in New York City by William Bratton in 1994, a person's skin color has, at least in part, become "the basis for distinguishing the undesirable from the desirable."

Concerns for equity aside, scholars are not even in agreement on the efficacy of broken windows policing. In [15], Wesley Skogan presents data from a 6-year study of six urban centers in the US supporting a link between civil disorder and crime. However, in [7], Bernard Harcourt points out deficiencies within Skogan's data and analysis, and Harcourt's re-analysis of the data refutes Skogan's initial findings. Also, in [8], Dan Kahan attributes the decline in crime in New York in the 1990s to Bratton's policing strategies, whereas Andrew Karmen argues in [9] that the decline in crime could be due to many causal factors that occurred in New York City in the 1990s (decline of drug use, lower unemployment rates, increased college enrollment, etc.). In an empirical study of 196 Chicago census tracts, Robert Sampson and Stephen Raudenbush showed that the "relationship between public disorder and crime is spurious," but that "collective efficacy" – defined as the "cohesion among residents" and "shared expectations for the social control of public spaces" – is linked to lower rates of violent crime [14]. In summary, although some have termed Kelling's and Wilson's original 1982 article the "bible of policing" [4], the broken windows theory has yet to be conclusively verified, due to the inherent difficulty in defining, isolating, and measuring the independent variable of "disorder."

To explore the two themes of the equity and efficacy of broken windows policing in New York City, we designed a semester-long statistics project which involves collecting, organizing, and analyzing data from New York City arraignment courts and the Uniform Crime Reporting Program of the Federal Bureau of Investigation.

2. POLICE REFORM ORGANIZING PROJECT

The Court Monitoring Project we implemented in our statistics class was a part of a larger effort of the same name first started by Robert Gangi and the Police Reform Organizing Project (PROP). Gangi, a lifelong New Yorker and advocate for criminal justice reform, founded PROP in 2011 after nearly 30 years as the director of the Correctional Association of New York. PROP's vision is to "[expose] discriminatory and abusive practices of the NYPD that routinely and disproportionately affect [New York City's] low-income communities and people of color" [16].

The Court Monitoring Project is an effort by PROP to collect data from New York City's misdemeanor arraignment courts by visiting and observing the various arraignment sessions to record the criminal charge, verdict, and perceived race, age, and gender of defendants who appear before the judge. This data is collected from the various criminal courts within the five boroughs of New York City and amassed in press releases to reveal common criminal charges levied by the New York Police Department (NYPD) and patterns in the types of individuals against whom such charges are levied [16]. Since all New Yorkers who are arrested or issued a summons must be arraigned before a judge, the theory behind the Court Monitoring Project is that observing patterns of charges and defendants within the courts will reveal on-the-ground practices of the NYPD. Furthermore, PROP focuses on low-level misdemeanor offenses such as loitering and vagrancy because an emphasis by the police to control such offenses is one indicator of a broken windows policing strategy. Figure 1 summarizes the main efforts of PROP's Court Monitoring Project.

3. PROJECT DESCRIPTION

In this section we describe in great detail the sequence of the semester-long project. The main student-learning outcomes of the project are three-fold:

1. Data collection: Students learn that data collection is a difficult process, and use sound data collection techniques to collect raw data from their environment.

COURT MONITORING PROJECT

observe

NYC'S ARRAIGNMENT AND SUMMONS COURTS

+

talk

WITH LAWYERS AND DEFENDANTS

gain

KNOWLEDGE AND FACTS ABOUT THE NYPD'S PRACTICES ON THE GROUND

PROP VISITS COURTS IN NYC TO MONITOR:

- THE MOST COMMON ARRESTS AND SUMMONSES MADE BY THE NYPD
- PATTERNS IN WHO IS BEING ARRESTED AND TICKETED
- COLLATERAL CONSEQUENCES OF NON-VIOLENT, LOW-LEVEL CHARGES

Figure 1. PROP's description of its Court Monitoring Project. Used with permission [16].

2. Data organization: Students organize raw data using spreadsheets and other statistical software, and construct visual representations appropriate to the type of data.
3. Data analysis: Students interpret data and use their interpretations to construct and/or support an argument.

The project was framed as an invitation for students to collect, organize, and analyze data in an effort to answer the following two "Big Questions," which were alluded to in Section 1.

Big Question #1: Is broken windows policing equitable?

Big Question #2: Is broken windows policing effective?

These questions are rightfully described as "big" due to the complexity of broken windows policing and the difficulty in defining and measuring equity and efficacy. To stimulate thought and debate within the class, we intentionally left the Big Questions vague and broad. However, to slightly uniformize the students' projects and to give them concrete statistical analysis to perform, we used the following "Little Questions," the answers to which help students argue a particular stance on the corresponding Big Questions.

Little Question #1: Are people of certain races over-represented as defendants within misdemeanor arraignment courts?

Little Question #2: Are misdemeanor arrest rates negatively correlated with violent crime rates?

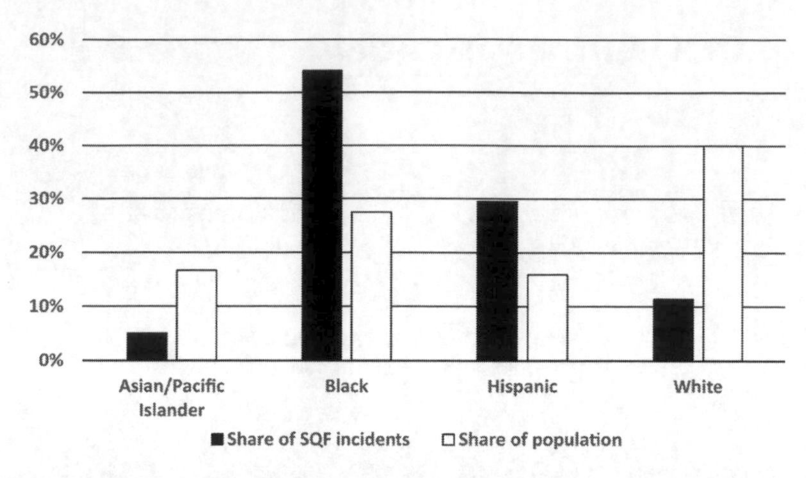

Figure 2. The 2015 New York City SQF incidents versus population (by race).

We will describe the sequence of the project in four stages: *motivation, orientation, data collection*, and *synthesis and analysis*.

3.1. Motivation

We built motivation for the project in two ways: a study of data from the NYPD's Stop, Question, and Frisk program, and a class visit from Robert Gangi.

The NYPD's Stop, Question, and Frisk (SQF) program is one means by which it has implemented a broken windows policing strategy. In the 1990s and 2000s, SQF and its practice in New York City was very controversial, due to the perceived influence of race on an officer's decision to stop a suspect [6]. After receiving a significant amount of negative media attention, the number of SQF incidents per year has decreased dramatically, from 685,724 in 2011 to 22,563 in 2015 [13]. For each incident, officers are required to fill out reports recording various variables such as the age, height, weight, race, and gender of the suspect, whether or not physical force was used, and whether or not a weapon or contraband was found.

In a class period early in the semester, students were given access to the SQF data for 2015, and asked to make various graphical representations of the data using the web-based statistical software StatCrunch. The purpose of the activity was to introduce students to the quantitative notion of "over-representation." Figure 2 shows a bar graph comparing the share of SQF incidents to the share of population by race (population data is from

Figure 3. A video that was shown in class which highlights Robert Gangi and the work of PROP. Used with permission [12].

the American Community Survey 2011–2015 (5-year estimates) [1, 2]. The over-representation of black and hispanic individuals in the SQF data led to a class discussion about possible causes for this over-representation, which include economic factors (poorer neighborhoods have higher crime rates and higher minority populations) and implicit biases within police officers. The idea for the SQF activity was inspired by work of L. Khadjavi [10].

Having developed the notion of over-representation in the context of the policing practice of SQF, we arranged a visit from Robert Gangi of PROP to discuss the theory of broken windows policing with the students. Figure 3 contains a video that was shown in class to introduce Mr. Gangi to the students. The video features Mr. Gangi and the work of PROP to raise awareness of discriminatory practices of the NYPD.

After the video, Mr. Gangi shared with the students his history in criminal justice reform, including his work to improve the conditions of pre-arraignment holding cells during his time with the Correctional Association of New York and his current efforts with PROP. He shared stories of interactions he has had with police officers, lawyers, and former police chief William Bratton illustrating the systemic pressures within the NYPD and New York City's criminal justice system that disproportionately affect low-income New Yorkers and New Yorkers of color. Mr. Gangi then shared the specifics behind PROP's Court Monitoring

Project and invited students to collect data to contribute to PROP's effort. His visit concluded with time for students to share stories about how policing has affected their own lives or the lives of their families, in both positive and negative ways.

3.2. Orientation

To know what the students would encounter in the courts, we attended an arraignment session with Mr. Gangi and other PROP representatives. The proceedings in the court moved very quickly to the untrained observer, especially when dealing with routine, low-level offenses. The majority of misdemeanor defendants spent less than 5 minutes standing before the judge to hear their criminal charges read and verdict announced. The two most common verdicts to misdemeanor offenses were "adjournment in contemplation of dismissal" (ACD) and a guilty plea for disorderly conduct. A verdict of ACD required defendants to stay out of legal trouble for a specified period of time in order for the charge to be dropped. In essence, an ACD verdict functioned as "not guilty." If an ACD was not granted, the judge often allowed defendants to plead guilty to a disorderly conduct violation (which is not a crime) which required defendants to pay a fine and/or complete community service, but granted them the benefit of not having a misdemeanor offense on their criminal record.

To prepare students for the pace, culture, and vernacular of this "conveyer-belt" justice system, we decided to use class time to inform them of what they should expect to see when visiting the courts. We also handed out data collection sheets to each student which asked them to record, to the best of their ability, the following variables for each defendant: race, age, gender, whether or not the defendant was in custody prior to the arraignment, criminal charges (code or title), verdict (ACD or guilty plea), fee/fine, days of community service, whether or not the defendant was released, and time appearing before the judge.

3.3. Data collection

The data collection process had two phases - one for each Little Question the project was designed to address. Raw data collected by students observing the courts provided evidence to help answer whether or not individuals of certain races are over-represented in New York City misdemeanor arraignment courts. Crime and arrest data collected from the Uniform Crime Reporting (UCR) Program of the Federal Bureau of Investigation provided evidence to help answer whether or not there is a relationship between misdemeanor arrest rates and violent crime.

Race	# of defendants	% of defendants	% of population
Asian	16	6%	17%
Black	136	54%	27%
Hispanic	74	29%	16%
White	27	11%	40%
Total	**253**	**100%**	**100%**

Figure 4. Our data on race from the Court Monitoring Project [22].

3.3.1. Community Days and data collection in the courts

Our institution, Guttman Community College, places great emphasis on social justice and service learning. These values are implemented on a broad level across the college through our practice of "Community Days." Community Days are two class days built into our academic calendar each semester during which all classes are canceled to give students the opportunity to volunteer with community organizations and complete field research for assignments. Faculty at Guttman carefully prescribe the Community Days experience for our first - year students. When we assigned the Court Monitoring Project to our first-year statistics students, we required they use one of their two Community Days to visit and observe a New York City misdemeanor arraignment court.

In preparation for Community Days, students signed up to attend a New York City arraignment court of their choice (most chose the court within their home borough). We arranged for some Guttman faculty, Mr. Gangi, and other PROP representatives to meet groups of students at the courts to help them collect data. Students were required to collect data in the court for 3 hours. After Community Days, the different student groups consolidated their data into one spreadsheet. All together our students collected a sample of 253 different defendants. The data on race can be seen in Figure 4.

3.3.2. Data collection using UCR

Students obtained access to the FBI's UCR data through the online research tool Social Explorer (Guttman provides its students with a professional license to Social Explorer). Students were assigned one of the 50 US states and were guided to use Social Explorer to obtain misdemeanor arrests rates, total violent and property crime, and violent and property crime rates for each county within their assigned state in 2010 and 2012 (the most recent UCR data available via Social Explorer).

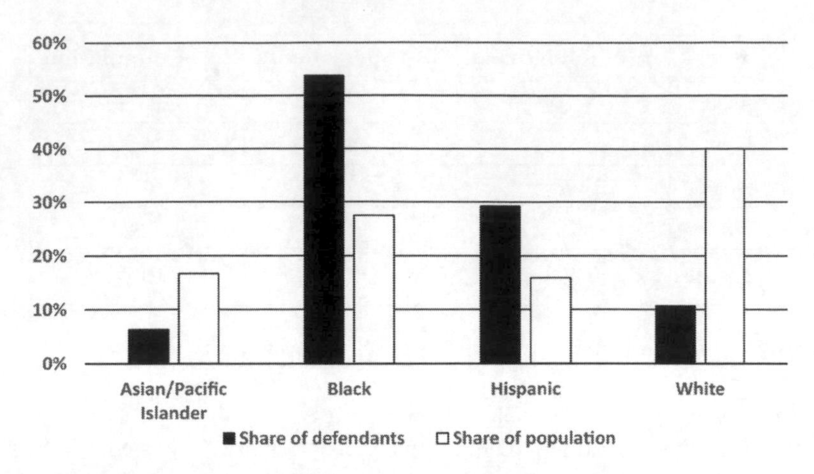

Figure 5. Misdemeanor defendants versus population (by race).

3.4. Synthesis and analysis

To present their findings, students were asked to write a report that included various graphical representations of their data along with interpretations of these graphs. A description of the various graphs follows.

3.4.1. A contingency table and bar chart displaying the race of our sample of 253 defendants

Students were asked to construct a contingency table and bar chart displaying the race of our sample of defendants. To answer Little Question #1, students had to compare the data from our sample of defendants [22] with data from the entire New York City population [1, 2]. Figures 4 and 5 show the contingency table and bar chart displaying race. The striking similarity between Figure 2 and Figure 5 is not a mistake: the proportions of races observed within our sample of misdemeanor defendants were nearly identical to the proportions of races from the 2015 SQF data.

After constructing this table and bar chart, students were equipped with statistical evidence to answer Little Question #1. In designing the project, we expected the data we would collect to provide a preponderance of evidence suggesting an affirmative answer to this question. As seen here, we were not surprised. Notice that this data is ripe for hypothesis testing (specifically chi-squared); however, this content was beyond the scope of our first semester course (at Guttman we stretch a standard introductory statistics course across two semesters for our students without proficiency in basic algebra. We did however make use of this data in the second semester course when treating inferential statistics).

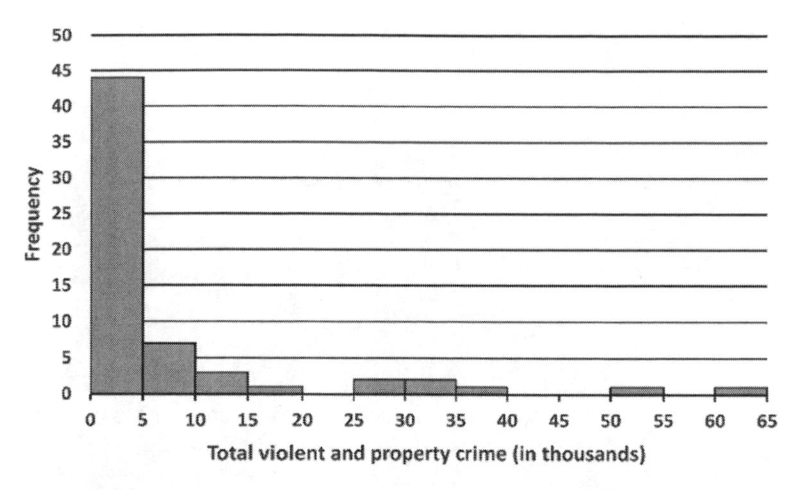

Figure 6. Histogram of total crime in New York by county (2012). Data from UCR [20].

3.4.2. Histograms displaying total violent and property crime and violent and property crime rates (by county)

Students were asked to construct two histograms displaying the numerical variables "total violent and property crime" and "violent and property crime rate." Each data point in their histograms represented a different county in the state they were assigned to study. Examples of these histograms for New York state are shown in Figures 6 and 7.

Most students' histograms looked similar to those in Figures 6 and 7, showing a heavy right skew for total crime and a rough symmetry for crime rate. Students were asked to explain the heavy skew for the histogram displaying total crime (most counties within a state have similar populations, and thus comparable total crime, whereas a few counties with large urban centers and large populations will be outliers). Students were also asked to explain what a crime rate is, and why the histogram showing crime rate is a better way to compare counties of different sizes. For instance, the two outliers in Figure 6 are Queens County (Queens) and Kings County (Brooklyn), the two most populous counties in New York State. However, these two counties fall right in the center of the histogram measuring crime rate. Students were also asked to describe and interpret the center and spread of each histogram using the appropriate measure given the shape. Median, mean, standard deviation, and interquartile range were all calculated using StatCrunch.

3.4.3 Boxplots displaying misdemeanor arrest rates

The project also asked students to make boxplots displaying misdemeanor arrest rates by county for the years 2010 and 2012. A sample of

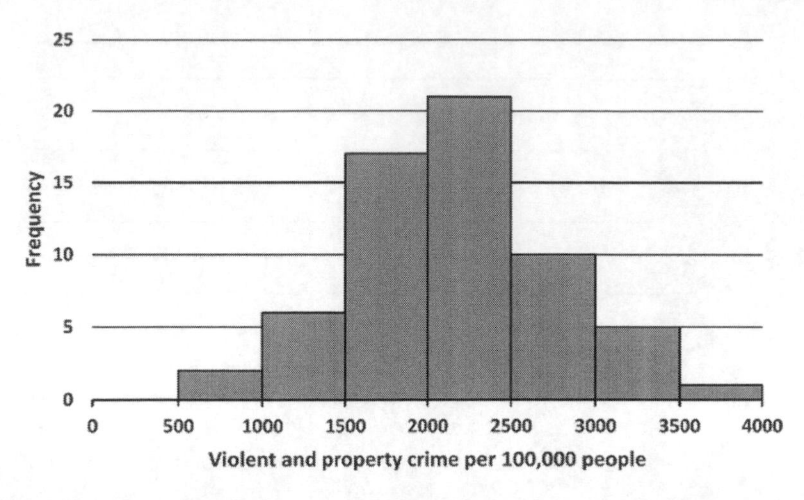

Figure 7. Histogram of crime rates in New York by county (2012). Data from UCR [20].

such a graph is shown in Figure 8 for New York state (the arrest data contains all arrest categories reported by the UCR except for the eight "Part I" offenses: homicide, rape, robbery, aggravated assault, burglary, larceny-theft, motor vehicle theft, and arson [21]). Students were asked to compare the two boxplots, and interpret the five-number summary in the context of misdemeanor arrest rates within the counties of their assigned states. Students were also asked to name any counties that appeared as potential outliers, and research those counties to offer a possible explanation for their extreme statistics. For example, in Figure 8, what appears to be a single outlier is in fact five different outliers, and due to the history of broken windows policing in New York City, it is not surprising that those five counties are precisely the five boroughs of New York City (Bronx, Brooklyn, Manhattan, Queens, and Staten Island).

3.4.4. A scatterplot comparing the misdemeanor arrest rates and violent and property crime rates (by county)

The scatter plot requirement of the project involved a very simple comparison of two variables. In essence, we took the rather naïve view that counties with higher misdemeanor arrest rates are more likely to be implementing some form of a broken windows policy. Plotting arrest rates against crime rates could then suggest whether or not broken windows policing is effective (a positive trend would suggest it is not, whereas a negative trend would suggest otherwise). Such a scatter plot for New York counties in 2012 is shown in Figure 9. The slope of the regression line is 0.24, and is significantly different from zero ($p = 0.003$).

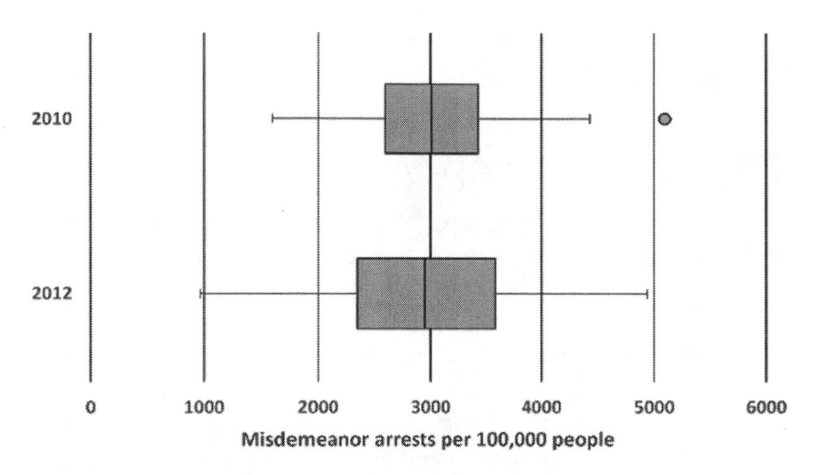

Figure 8. Boxplots of misdemeanor arrest rates in New York by county. Data from UCR [17, 18].

Students constructed similar scatter plots for their assigned states, reported the correlation coefficient and the regression line, and stated the contextual significance of the slope and y-intercept of the regression line. They also used the regression line to predict the crime rate in a county with an arrest rate of 3,000.

Students' scatter plots varied in strength and trend, suggesting that Little Question #2 is not easily answered. This was not surprising, due to our overly-simplistic analysis of a very complex issue, and the controversy surrounding the question of the efficacy of broken windows policing within both political and academic spheres (see Section 1).

We recognize many shortcomings and oversimplifications in using Little Question #2 as evidence for answering Big Question #2. First, using the misdemeanor arrest rate as the sole indicator of a county's participation in broken windows policing is too simple. Second, if broken windows policing is effective and reducing social disorder through misdemeanor arrests does indeed help reduce serious crime, we might not expect to see a reduced crime rate in the same year we see an increased arrest rate. In other words, there might be a time delay between cause and effect. Thus, to graph 2012 arrest rates against 2012 crime rates is again overly simplistic. Nevertheless, since the project was already fairly lengthy and involved, we decided to settle on this basic approach in the actual implementation with students.

However, in the writing of this article, we refined the crude approach to exploring the efficacy of broken windows policing that we used in the class project. First, we suggest that the ratio of a county's misdemeanor

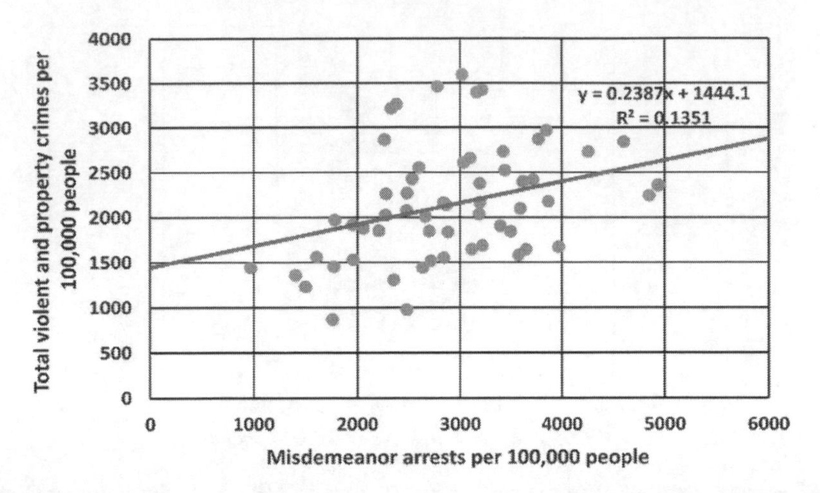

Figure 9. Misdemeanor arrest rate versus crime rate in New York counties (2012). Data from UCR [18, 20].

arrests to total incidents of violent and property crime is a better indicator of that county's participation in broken windows policing than misdemeanor arrest rate alone (a high ratio indicates that a county is emphasizing the control of misdemeanor crime *in relation to* the level of violent crime present within the county). We will call this ratio the "broken windows indicator." Second, to account for time delay, we will plot the 2010 broken windows indicators for each county against the 2-year percentage change in violent and property crime between 2010 and 2012 (one could explore a similar plot using the changes across a 1-year span, or a 3-year span, etc.).

Figure 10 illustrates this slightly more sophisticated approach to study broken windows' efficacy. The scatter plot is not clearly linear, and it has an influential point, but we include the regression line to discuss the extent to which it appropriately models the data. The positive slope of the regression line is significantly different from zero ($p = 0.0004$). If the influential point is removed, the slope of the regression line is 5.26 and remains significantly different from zero ($p = 0.04$). The linear model thus indicates that the greater a county subscribed to a broken windows policing theory in 2010 (as measured by our indicator), the greater their 2-year percentage change in crime was between 2010 and 2012. This refined (yet still over-simplified) analysis does not support the conclusion that broken windows policing is effective in curtailing violent crime. If we were to do this project again, we would make space in class to discuss how one might quantify a county's participation in broken windows policing, perhaps landing on something similar to our broken windows indicator. We would also brainstorm with the class what other data (in

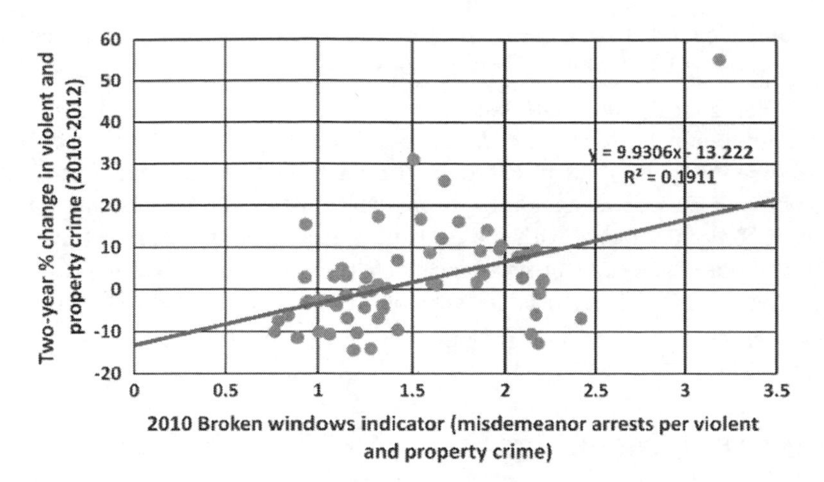

Figure 10. Broken windows indicator versus % change in crime. Data from UCR [17–20].

addition to crime and arrest data) might be helpful in determining the extent to which a particular police department is using a broken windows policing strategy. The depth of quantitative literacy and proportional reasoning a student would need to fully understand (and critique) Figure 10 might take a few classes to develop, but would likely lead to interesting discussion.

4. STUDENT RESPONSE

Anecdotally, student response to the project was very positive. Many students came back from the courts sharing that they were surprised by how interesting it could be to sit in a court room for 3 hours. The presence of one group of students taking notes in an arraignment court was noticeable enough to prompt the judge to address the students in private to encourage them in their studies! Multiple students referenced the project in their course evaluations. Here are a few direct quotes:

> The most impactful experience I had in the course was going to the court house and recording data on my own.

> I don't like math but using statistics for the Court Monitoring Project was interesting.

> The Court Monitoring Project made the most impact on me. It did because I learned that broken windows really targets people of color by

attending court and seeing it with my own eyes. Also, I applied everything I learned in this semester in this project.

Being mostly people of color, our students had likely already formed (whether subconciously or conciously) the hypothesis that race has become, in part, "the basis for distinguishing the desirable from the undesirable" as Kelling and Wilson feared it would [23]. This project equipped our students with the tools necessary to explore the veracity of this hypothesis using statistical means. Students were not at all surprised by the findings of our class's data collection in the courts, but that the data agreed with what they already suspected was affirming of their experiences and worldviews.

5. NOTES FOR EDUCATORS

The description of the project in this article is admittedly catered to the Guttman's geographical and institutional context in New York City. We conclude with some suggestions for instructors who are interested in implementing this project within their own settings.

5.1. Data collection/Data sources

Guttman's practice of Community Days largely facilitated the data collection process in the courts. Since classes were canceled for the specified purpose of allowing students to engage in field research for their classes, we felt the freedom to use Community Days for data collection. If the instructor is hesitant to ask students to collect their own data from the courts (which represents a large time commitment outside of class), we suggest the use of data from the United States Sentencing Commission (USSC) [3]. The USSC publishes demographic and legal data for individual offenders. These data sets are SAS and SPSS compatible, and are fairly large (~50,000 observations with ~10,000 variables). We recommend the following list of variables for study, though we invite interested instructors to peruse the codebook for other relevant variables:

- AGE - offender's age
- FINE - dollar amount of fine ordered
- MONRACE - offender's race
- MONSEX - offender's sex
- OFFTYPSB - primary offense type
- PRESENT - pre-sentence detention status
- SENTIMP - sentence type (fine only, prison, probation, etc.)
- SENTTOT0 - prison sentence length (in months)

For those instructor's who *can* make time for students to visit the courts outside of class (perhaps by canceling a class section) we recommend organizing students in small groups to attend the arraignment sessions and coordinating different groups to observe different courts. If only one court is geographically feasible, due to the location of the instructor's institution, different groups should attend at different times or different days so that the class as a whole can obtain a larger sample. The instructor should provide the students with a table whose columns are headed with a list of variables students should try to record for each arraignment. If the instructor's institution is in a fairly monoracial locale, then students could be encouraged to record visually perceivable demographic identifiers other than race (such as age or sex). Little Question #1 would have to be appropriately adjusted to study these identifiers (e.g., "Are people of certain ages over-represented as defendants within misdemeanor arraignment courts?").

Guttman provides students with a professional license for Social Explorer which allowed our classes to access the UCR data. At an institution without such access, an instructor could make use of the following free data sets to find demographic, crime, and arrest data:

- The UCR data tool [21] is a free resource that provides violent and property crime data. The tool is flexible enough to report data at the city, county, and state levels.
- Each year the FBI's National Incident-Based Reporting System (NIBRS) publishes the number of occurrences of over 50 types of criminal offenses. Again, these data are available at the city, county, and state levels (in Excel spreadsheets and a convenient map tool). As of the publishing of this article, only about one-third of law enforcement agencies within the U.S.A. participate in NIBRS, but the FBI has made "nationwide implementation of the NIBRS a top priority" [11].
- The United States Census Bureau's American FactFinder [5] is a great tool for finding general demographic data at the city, county, zip code, or state level. An instructor could make use of this tool to compare population demographics with the demographics of the sample of individuals observed in the misdemeanor arraignment courts.

5.2. Partner with a local police reform organization

We were so fortunate to have Mr. Gangi's support and cooperation, and partnering with PROP gave the project life and a sense of purpose that extended beyond the classroom walls. We eventually shared the data we collected with PROP to support its larger Court Monitoring Project. We encourage interested instructors to make a few inquiries (phone calls, emails) to police reform organizations within their vicinities. Our partnership with PROP began with a Google search and an email.

5.3. Logistical and assessment information

In our implementation of the project, we had students attend the courts in groups, but each student completed an individual report (for Little Question #2, each student was assigned a different state to study). The project was worth 25% of a student's final grade.

The sheer length of the project required a great degree of scaffolding. The students were required to: (i) collect UCR data from Social Explorer; (ii) construct six different figures (one table and five graphs); (iii) interpret each; and then (iv) write a report summarizing their findings. Each of these four steps in the process had its own due date and the completion of each part by its due date was worth a certain amount of points towards the project's final score. These scaffolded due dates helped ensure students were making progress towards the final product and gave us the opportunity to give feedback along the way. Furthermore, we made regular space in class for students to work on their project with the help of the instructor and their peers to help supplement the amount of time required outside of class to complete it.

ACKNOWLEDGEMENTS

We would like to acknowledge Lily Khadjavi for inspiring our project design and topic through her public work with mathematics and social justice and through private correspondence. The idea to find partnership with a community organization was suggested by Dave Kung and Forest Fisher (independently). We are grateful to both Paul Notice and Robert Gangi for allowing us to use their videos and images in this article, and especially to Mr. Gangi for engaging our students, supporting the project, and advocating for New Yorkers who are powerless to do so for themselves. Liang Zhao was a willing and enthusiastic partner in implementing this project with our students. The referees and guest editors provided very helpful comments to sharpen the content of this article. Finally, we are grateful to our students who worked very hard to complete this fairly lengthy and complex project.

REFERENCES

1. American Community Survey (5-Year Estimates). Race, 2015. Prepared by Social Explorer.
2. American Community Survey (5-Year Estimates). Hispanic or Latino by race, 2015. Prepared by Social Explorer.
3. Commission Datafiles. 2017, April 07. http://www.ussc.gov/research/datafiles/commission-datafiles#individual. Accessed 10 April 2017.
4. Cullen, K. and G. Staff. 1997, May 25. The commish. http://www.highbeam.com/doc/1P2-8424197.html?refid=easy_hf. Accessed 13 April 2017.
5. Data Access and Dissemination Systems (DADS). 2010, October 05. American FactFinder. https://factfinder.census.gov/faces/nav/jsf/pages/index.xhtml. Accessed 17 October 2017.
6. Fagan, J. and G. Davies. 2000. Street stops and broken windows: Terry, race, and disorder in New York City. *Fordham Urban Law Journal*. 28: 457–504.
7. Harcourt, B. 1998. Reflecting on the subject: A critique of the social influence conception of deterrence, the broken windows theory, and order-maintenance policing New York style. *Michigan Law Review*. 7(2): 291–389.
8. Kahan, D. M. 1997. Response: Between economics and sociology: The new path of deterrence. (response to article by Neal Kumar Katyal in this issue, p. 2385). *Michigan Law Review*. 95(8): 2477–2497.
9. Karmen, A. and Project Muse. 2000. *New York Murder Mystery: The True Story Behind the Crime Crash of the 1990s*. New York: New York University Press.
10. Khadjavi, L. S. 2006. Driving while black in the city of angels. *Chance*. 19(2): 43–46. doi:10.1080/09332480.2006.10722786
11. National Incident-Based Reporting System. 2016, April 20. https://ucr.fbi.gov/nibrs-overview. Accessed 17 October 2017.
12. Notice, P. (Producer). 2016, February 11. The impact of "broken windows" policing on black communities [Video file]. http://elitedaily.com/news/impact-brokenwindows-policing-black-communities/1380144/. Accessed 10 April 2017.
13. NYPD Stop Question and Frisk Report Data Base. n.d. http://www.nyc.gov/html/nypd/html/analysis_and_planning/stop_question_and_frisk_report.shtml. Accessed 10 April 2017.
14. Sampson, R. and S. Raudenbush. 1999. Systematic social observation of public spaces: A new look at disorder in urban neighborhoods 1. *American Journal of Sociology*. 105(3): 603–651.
15. Skogan, W. 1992. *Disorder and Decline: Crime and the Spiral of Decay in American Neighborhoods*. Berkeley, CA: University of California Press.
16. The Court Monitoring Project. n.d. http://www.policereformorganizing-project.org/court-monitoring-project/. Accessed 10 April 2017.
17. Uniform Crime Reporting Crime Data. Arrests, 2010. Prepared by Social Explorer.

18. Uniform Crime Reporting Crime Data. Arrests, 2012. Prepared by Social Explorer.
19. Uniform Crime Reporting Crime Data. Violent and property crimes reported (rate per 100,000 population), 2010. Prepared by Social Explorer.
20. Uniform Crime Reporting Crime Data. Violent and property crimes reported (rate per 100,000 population), 2012. Prepared by Social Explorer.
21. Uniform Crime Reporting Statistics. n.d. https://www.ucrdatatool.gov/.
22. Warner, J. and L. Zhao. 2016, November 14. [Court Monitoring Project Data]. Unpublished raw data. Data collected by students across three sections of introductory statistics. Available at http://www.statcrunch.com/app/index.php?dataid=1999330. Accessed 10 April 2017.
23. Wilson, J. Q. and G. L. Kelling. 1982. Broken windows: The police and neighborhood safety. *The Atlantic*. 249(3): 29–38.

Meaningful Mathematics: A Social-Justice-Themed-Introductory Statistics Course

jenn berg, Catherine A. Buell, Danette Day, and Rhonda Evans

Abstract: As an interdisciplinary team, we set out to create an applied statistics course that would cover the traditional introductory statistics topics in a consistent framework of social justice. The goal was to motivate students to understand and learn math while deepening their understanding of the interplay, at local and global levels, between social and economic issues: for example, crime, victimization, political access, wealth, education, health, gender, and race. This paper describes the process of creating the course, the pedagogical decisions, the intentionality in topic presentation, samples of classwork, and student impressions of the semester-long course.

1. INTRODUCTION AND OVERVIEW

Introductory statistics is a common course taken by first-year students at colleges and universities and is tasked with serving many needs across campus: fulfilling a general education requirement; serving as a course prerequisite to research design and analysis; or as an option for first-year students who have a weak algebraic background. As statistics is prevalent outside academia, some schools provide statistics courses particular to given majors, such as business, social sciences, or sociology; this places the mathematical topics in a familiar setting. Schools that do not have the resources to do so may find their statistics courses lack a context that seems relevant to the students. An emerging trend in higher education is to select broad themes for courses, which creates a rich contextual environment in which students can develop skills. Our

Color versions of one or more of the figures in this article can be found online at www.tandfonline.com/upri.

increasing understanding of the role emotional engagement plays in capturing and maintaining student attention, and the follow-up consequences for long-term retention of knowledge suggest that selecting an emotionally engaging theme for such courses allows non-discipline-specific general education courses to engender increased learning, due to increased student engagement and a sense of the subject's relevance.

It is in this light, as well in light of the increased public discourse around issues of social justice, that we decided to develop a social-justice-themed version of the applied statistics course at our institution. Although the topic clearly provided an emotionally engaging context, in order to leverage the advantages of teaching mathematical concepts in a consistent context, we needed to develop course materials that addressed all the mathematical topics in the core syllabus. This work builds off efforts of others [1, 4] to offer methods for including lessons that address mathematical content and issues of social justice. We decided to have the entire course have a social justice theme. The primary reason was the desire to provide a consistent context; however, a close secondary reason was to honor the emotional weight of the topics and avoid the dissonance of asking students to grapple with issues of inequity and violence in their communities on one day, but return to "neutral" topics on the next. In the following sections we describe the planning and course design, present a variety of student activity materials and how these examples were received by the students in the course, and finally review pre- and post-survey data from students enrolled in the two sections.

2. COURSE DESIGN

As our goal was to create a statistics course fully integrated with social justice, we partnered with professors in education and sociology, and we received support from the Center for Teaching and Learning for a pedagogical discussion group that would meet bi-weekly throughout the spring term. At the onset of this project, all members of our team shared a commitment to improving student learning outcomes through engaged learning, as well as a common concern for social justice and a belief that empirical insights on such issues can supply our students with the knowledge to become responsible citizens.

This team proposed to select readings, find data, develop student learning objectives, develop student learning assessments, and construct university policy recommendations for a social-justice-themed applied statistics course. The project aimed to foster student's ethical reasoning in one of the most commonly taken MATH classes. We established course goals for the mathematical content, social justice content, and course overall. For the course overall, we wanted students to have the ability to seek, engage, and be informed by multiple perspectives and to have the ability to use critical inquiry and quantitative reasoning to identify a problem, research solutions, analyze results, evaluate choices, and make decisions. These goals were reflected in the desired outcomes for both the mathematical and social justice content.

Mathematical content:

- Understand and create statistical data numerically, graphically, through formulas, and verbally.
- Specific topics include: common statistical vocabulary, visual representations of data, measures of central tendency, measures of variation, probability, p-values, confidence intervals, regression, t-tests, pattern description.

Social justice content:

- Recognize the power of math as an essential analytical tool to understand the world, and recognize their own power as active citizens in building a democratic, equitable society.
- Become more motivated to understand and learn math while deepening their understanding of the interplay of social and economic issues, e.g., crime, victimization, political access, wealth, education, health, gender, and race.

In addition, we identified several programs of study whose curriculum could benefit from the social-justice-themed applied statistics: Criminal Justice, History, Political Science, Human Services, Police Certification, Sociology, Secondary Education, Middle School Education, Nursing, Psychology, and Environment Science. We reached out to these departments and many recommended their students take the course through formal and informal advising. We scheduled two sections of the social-justice-themed applied statistics course during the semester with two different professors, which allowed for the sharing of activities, syllabi, and classroom experiences. Homework assignments were done through an on-line system. We chose to use projects and weekly (or bi-weekly) quizzes throughout the semester and a final exam.

After two of the authors attended a workshop on intertwining mathematics and social justice course design, we realized that without discussing issues of allocation of resources we were not providing students with the background to make inferential connections in the material when race, gender, ethnicity, and other difficult topics of privilege and inequity are presented. Without this context, the students who lack background knowledge could easily fall back on pre-established stereotypes and anecdotal evidence and fail to make well-rounded, informed connections. Therefore, it was important to order the social justice topics carefully. Figure 1 shows the ordering we kept in mind when designing activities for the course. We wanted to begin with chances for students to unpack their personal beliefs and dispel stereotypes as a way to provide a consistent socially aware background on which to build.

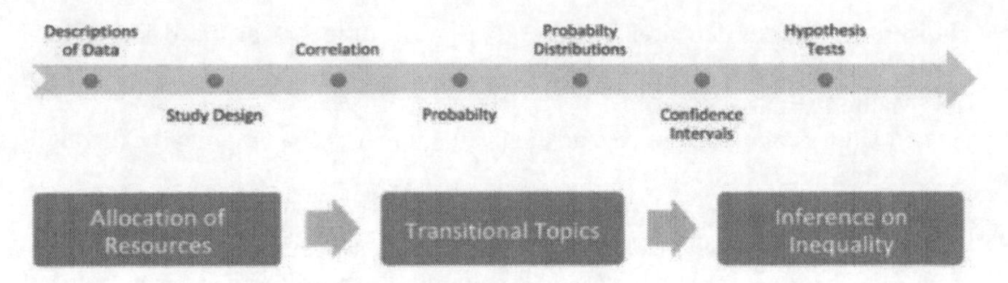

Figure 1. Ordering of the course.

Hence, early in the semester we would contextualize the statistical concepts from what we deemed *Allocation of Resources*: access to heath care, wealth, income, access to early childhood education, taxation, access to healthy food, cost of college, air quality, proximity to major highways/housing, access to transportation, dangerous occupations. As students developed their understanding of resource allocation, confronting some priorly-held beliefs about equality in the United States, we would begin to introduce *Transitional Topics*: infant mortality, preterm births, drug use, sentencing, political participation, age of retirement, unemployment/underemployment, consequences of war, policing; these topics were not highly controversial, but still called upon the knowledge of how allocation of resources plays a role. Later on students could begin to use various statistical tools in more nuanced social justice contexts, making *Inference on Inequality*: education gap (race/gender/ethnicity), imprisonment/jail/executions, crime, victimization, pay gap, mass incarceration. This scaffolding, which had students considering issues of social justice issues in complex ways, led to student interaction with topics they may have had prior prejudices or deeply emotional experiences.

3. ACTIVITIES

In the following section, we present examples of activities that were used throughout the semester and a sample project. We describe where the activity landed in the semester, which social justice topics were addressed, and in some cases we discuss student experiences with the material. Exams and quizzes used questions similar to those from the in-class activities.

3.1. Setting the stage

We performed several crucial and intentional activities in the first week of class to 'set the scene' for this course. First, the students wrote a mathematical autobiography about their experiences in math classes and their academic achievement and background. Then the professor held one-on-one meetings with the students to discuss the autobiographies and the professor's teaching philosophy. Finally, through conversation and example, there were daily efforts to create and cultivate an environment where students were both willing to be disturbed [3] and to have respectful, data-driven conversations and debates.

3.2. First-day activity

We felt that it was important that early in the course students gained an understanding of the sort of topics we would be discussing in the semester, so that they could decide if the course would be appropriate for them. If not, they could change enrollment into one of the sections of the course without a social justice theme. For a first day activity, we selected numerous examples of questions from statistics texts which were not focused on social justice. We asked students in groups to review each of the problems, decide what topic(s) they thought the questions were covering, and create a table summarizing their results. After each group completed their table, the instructor compiled the results of all the groups on the board and facilitated a class discussion. This activity set the stage for some of the first mathematical concepts the course would discuss, such as data type, variables, and organizing data. The course discussion highlighted that mathematics courses usually attempt to cover "safe" topics, such as commerce, entertainment, and science whereas topics such as race, poverty, and crime are typically avoided. (Please contact either of the authors for this activity.)

3.3. Graphical displays of wealth distribution

This activity was inspired by an article by Paul Solman featured on PBS [2] that showed the wealth distributions for the quintiles of the population of the US, Sweden, and "Freelandia."

In our activity, students were given two blank circles and asked to draw the percentage of wealth they believed each quintile of the population held in the US and Sweden and explain their reasoning. This activity can be found in full as part of the online appendix of this issue.

These questions framed the follow-up discussion of what students know, or do not know, about wealth distribution in the US and other countries. A majority of students had a larger gap between the wealthiest group and the lower quintiles in Sweden than in the US. After the activity, students were shown the actual distributions and had additional prompts to both uncover what preconceptions exists in each situation and why and understand what these distributions meant.

3.4. Graphical displays of data: Creating effective histograms

We asked students to use data management software (such as Excel or Google Sheets) to create graphical representations of data (our student email service is on Google, and hence students had access to Google Drive products and sharing as part of their school email account). Using a table of the percent of children who live in poverty in each state, we asked students to make a histogram of the data. There was wide variety in the quality of the histograms, so one in-class activity had students look at samples of their peers' work to reflect on what attributes make for a good graphical representation. After selecting the histograms that best displayed the data from the table, the class was able to discuss some of the implications of the data. In this way, we were able to reinforce basic ideas such as titles, axis labels, and accuracy of graphs, while exposing students to data that would help them better understand the reality of wealth distribution in the US. The worksheets are available in the on-line appendix of this *PRIMUS* issue. The data set is also available in the online appendix.

3.5. Graphical displays of data: Box-plot pre-work

To bridge students' work outside of class to the work done in the class, the instructors regularly asked students to answer a few questions on the day's topic and have those answers available at the start of class time (completion of this work was part of the participation portion of the course grade). For example, in the section of the text connecting numerical measures of data and graphical displays of data, students were asked to look over a graph exploring traffic density and income groups of Sao Paulo, Brazil. This activity can be found in the on-line appendix. Here students had to make sense of the graph using labels and interpret what the six box-plots meant. Class time opened with student discussion in pairs. Students had difficulty interpreting the graphs before class discussion, but were able to use the shared context to develop understanding of the graphs throughout the day's discussion. During the class students were exposed to the methods researchers use to discuss income and explore how traffic density in Brazil compares to traffic density in the United States.

3.6. Numerical, graphical, and verbal descriptions of data

The selection of data presented early in the semester allowed the instructors to use the same data in multiple assignments. For example, a table from the

2010 census on the percentage of workers in high-risk occupations was used throughout the numerical and graphical descriptions of sections of the class. This table was used to help students understand how to read information from tables, and how to communicate data both through graphics and through text. Students were asked to give a written summary of some of the data that they found interesting.

After students had submitted their written summaries, select responses were shared with the whole class, and students were asked to select preferred summary sentences and create graphs that would illustrate the same point. This activity can be found in full in the on-line appendix.

3.7. Probability using jury bias and Stop and Frisk

The topics of 'policing' was part of the "transitional ideas" in social justice that paired nicely with developing student understanding of percentages, conditional probability, and probability rules. In addition, in these activities students had to use the quantitative results to make arguments and conclusions.

For jury bias, students were presented with the background of the Jena Six incident where six black students in central Louisiana were arrested after a fight that was sparked by racial tensions at the high school. One of the Jena Six had an all-white jury for his trial. The worksheet provided population data for the parish and information on jury selection including the attorneys' power of final selection.

Students were asked about the likelihood of an all-white jury and aspects of a "fair" trial. The main student inquiries included: Is it possible that this number is correct? What is the probability of at least one black juror? Why would the racial make-up of the jury matter? Where in the process did the selection fail? Student discussion on these questions highlighted the fact that it was important for the instructor to provide the background of allocation of resources, institutionalized racism, and privilege earlier in the semester, as these ideas were incorporated into the dialogue.

Also included with probability were Stop and Frisk data from the New York Police Department (NYPD). Students were provided with the racial and ethnic population breakdown of New York City and a breakdown by race for the number times the NYPD stopped, arrested, used force, found contraband, or found a weapon in 2009 under the Stop and Frisk program.

The probabilities that students were asked to calculate revolved around finding inequities in the stops and to address misconceptions with the program. Ultimately, students were asked why they thought this this practice was ruled unconstitutional.

Upon completing this exercise, a student said that they were provided biased data. Why had we selected the data to show a higher incidence of minority stops? We should have given them the data where the percentage of the stops matched the racial and ethnic percentages. The entire class launched into a discussion about where the data came from, who collected it, what the data showed, and how this student's request demonstrated the point of the exercise. The data, activity, and follow-up questions can be found in the on-line appendix.

3.8. Correlation and gun ownership

Students were presented with a table containing data on gun ownership, state gun law restriction rankings, gun deaths, and gun murders for each US state. The students were asked to use Excel or Google Sheets to create various scatterplots and to compute correlation coefficients for several pairs of variables. The students had to write a report to a legislator based on their findings. Students had to decide which variables to consider, decide how to interpret the results, and describe the data (outliers, meaning or variables, correlation). The table of data and the assignment are in the on-line appendix for this issue.

3.9. Experimental design project

Students were presented with a research topic:

> Researchers have suggested that when students from disadvantaged backgrounds receive peer mentoring during the freshman year of college it can improve their educational outcomes. You would like to test this claim and wonder if this matters more for male or female students. How can you set up an experiment to check this claim?

In class and in small groups, students were asked to develop the experiment, then given a rubric and asked to evaluate the proposal of other groups. After getting feedback from peers, students were able to refine their proposal and submit a final draft for credit toward the project portion of their course grade. Then the students (individually) were given a data set (one of two) that could have been the results of such an experiment with data on 100 "participants" including gender, hours spent in tutoring, and semester grade-point average. Students were then asked to use this data to answer the research question and draft a brief report of their results. The prompt and assignment was in the on-line appendix of this *PRIMUS* issue.

3.10. Quiz and final exam questions

As part of the course students were given bi-weekly quizzes on the mathematical concepts; when questions were presented with a context, we used a social justice context. For example, the quizzes on graphical displays of data asked questions about gender of doctoral degree recipients in different disciplines, or comparing national spending, as presented in pie charts. The quizzes on probability used the topic of juries, this time considering peremptory strikes and race, as well as sentencing in death row cases when the race of both the convict and victim was considered. Similarly, the cumulative final had questions that addressed wealth distribution, the proportion of population in prison disaggregated by race, median income as it compared to education level of a county, and comparison of sentences given for drug offenses involving crack or powder cocaine. Sometimes writing questions in a social justice context that were simple enough to be included on an in-class quiz required the instructors to simplify available raw data sets.

For example, the final included a question on a representative tri-city council which used the actual data from three towns in our university region, whereas a question that asked students to sketch a scatter plot comparing per student expenditure to standardized tests scores in the state pulled only small portion of the available data. We are willing to share any assessment materials with interested readers.

4. SURVEY RESULTS

As a part of the original course planning the team developed surveys to administer early in the term and near the end of the term. Common to the pre- and post-course survey were around 30 Likert scale questions designed to gauge student exposure to, and understanding of, a variety of social justice issues. The post-course survey also asked students about their experience in the course and gathered some demographic information. Students took the pre-course survey during the first week's supplemental instruction meeting, and took the post-course survey during the last week's supplemental instructional meeting. Due to withdrawals from the course and low participation from one of the sections the post-course survey had half as many responses. There were 66 respondents for the pre-course survey, and 33 respondents for the post-course survey. The survey was not set up to track changes in an individual student's responses.

Keeping these limitations in mind, we did an analysis of the comparison of student responses on the Likert scale questions. There were notable changes in the average responses of the class from the pre- to post-course survey. On average students were more likely to disagree with the statement "In the US, people in general are appropriately and consistently punished when they violate laws" (pre-survey 15 students agreed, post-survey six students agreed) and the statement "People with the same education, skills, and talents are equally likely to be hired" (pre-survey 12 students agreed, post-survey three agreed). Students were more likely to agree with the statement "Various sources of evidence inform my beliefs" (pre-survey four disagreed, post-survey two disagreed) and much more likely to agree with the statement “Statistical evidence is enough to change our beliefs on social justice issues” (pre-survey eight disagreed, post-survey one disagreed). Although the survey set-up limited the usefulness of these results, the responses suggest the course met some of the goals we had established: recognizing mathematics as a tool in understanding the world, and fostering a deeper understanding of the interplay between social and economic issues.

More powerful were the responses on the post-course survey asking students to reflect on the course. When asked “Do you feel that using social justice issues as a theme for your study of statistics help you learn, understand, and retain mathematical knowledge?” almost 80% (24 of 33) of the respondents said yes. Follow-up short answer replies (which were not common) were similar to the following:

> Math to me has always seemed somewhat pointless. I know that I would never realy [sic] need to know anything other than basic math skills in my everyday life. However, when a real life issues are applied, I found myself becoming more interested and the information was easier to retain.

Around half of the respondents (16 of 31) said that they felt their views on issues of social justice changed as a result of studying such issues in the context of statistics (again, with few written comments, but most of those responses suggested that the student was more aware of issues of social justice as a result of the course.) When asked if they would recommend the course to a peer, 87% of the respondents (27 of 31) said yes, and the no responses gave justifications along the lines of:

> It is tough to comprehend because it is not traditional math and it’s more about the why answer than just the finding the number,

which we have understood as a hidden positive response to the question.

5. CONCLUSION

The course was a success. The team that developed the course enjoyed the interactions outside of their home department and each member learned new things that were folded back in to their own instructional practice. The bi-weekly meetings in the semester before the course ran were often the highlight of week and deepened a sense of community for each of the members.

For mathematicians considering developing a similar course we recommend building a similar team or community of educators in sociology, the education department, or with community organizers in your area as well as finding a partner within the mathematics department. The new perspective on issues of social justice and how to address such issues in the college classroom was critical in the success of the course. Similarly important was the day-to-day development of course materials and regular conversations between the two instructors. Although the survey response numbers dropped between pre-course and post-course administration, the feedback was overwhelmingly positive, as was the feedback from advisors of students who took the class.

We would also strongly recommend using the social justice context for the entirety of the course, and not selecting only a few topics to sprinkle through the term. The cognitive advantage of using a consistent context is a good foundational reason as is the consistency of the emotional engagement and emotional energy students may have to expend in order to grapple with questions around the inequity in their communities. Although these reasons suffice for the choice to include social justice topics in each class meeting, we found that the careful choice in the order of social justice issues to discuss to be the most powerful reason to go all in. Without providing the shared understanding of how resources are, and have been, distributed in our society and without connecting the ideas of resource allocation to emerging issues of inequity, we suspect we would have encountered many more instances of student prejudice in classroom discussions when asking students to consider questions of inference. Furthermore, the regular exposure to topics of social justice allowed students to develop views, consider ways to use mathematics to explore issues of social justice, and build the skills necessary to understand and improve their society.

FUNDING

The course development group was funded by a Pedagogical Discussion Grant from the Fitchburg State University Center for Teaching and Learning. We are also very grateful to the organizers and mentors of the Workshop on Mathematics and Social Justice, funded by the Association of Colleges of the South and Fitchburg State University, in particular Lily Khadavi, Nathan Alexander, Joe Boltz, and Bonnie Shulman.

ORCID

Catherine A. Buell http://orcid.org/0000-0002-5716-2110
Danette Day http://orcid.org/0000-0003-3992-051X

REFERENCES

1. Gutstein, E., and Peterson, B. (Eds). 2005. *Rethinking Mathematics: Teaching Social Justice by the Numbers*, Milwaukee, WI: Rethinking Schools, Ltd.

2. Solman, P. 2011. *How does the US Slice the pie?*, PBS NewsHour, https://www.pbs.org/newshour/arts/easy-as-pie-inequality-in-downloadable-charts. Aired 12 August 2011.
3. Wheatley, M. J., 2002. *Turning To Each Other: Simple Conversations to Restore Hope to the Future*. San Francisco, CA: Berrett-Koshler Publishers, Inc.
4. Yang, K. W. Radical math. http://www.radicalmath.org/ Accessed October 2016.

Unnatural Disasters: Two Calculus Projects for Instructors Teaching Mathematics for Social Justice

Gizem Karaali and Lily S. Khadjavi

Abstract: We provide context and motivation for an instructor to use real-life examples in the calculus classroom. To this end we describe two specific project ideas, one related to the devastating impact of methylmercury fungicide in a grain seed supply and the other to a catastrophic methane leak. By using calculus in contexts that have social justice implications, we hope to empower students to reason for themselves, to use mathematics as a powerful tool to deepen their understanding of the world, and ultimately, to effectively confront the challenges society faces.

1. INTRODUCTION

Calculus is required for virtually all modern science and engineering majors. Even so, the vast majority of textbook problems remain grounded in artificially simple settings that do not force students to confront the very real-world fields they are studying. Generally speaking, calculus textbooks tend to, for example, prioritize modeling the trajectories of projectiles such as golf balls over studying applications with implications for the health of communities. Even those calculus texts that have been developed through an effort to model real-world scenarios may shy away from the most distressing ones; an exception is discussed below.

Our academic priorities, as reflected in the problems we have students work on, send a subtle, or perhaps not-so-subtle, signal about our values. Indeed, mathematics historian Judith Grabiner has noted on numerous occasions that mathematics in a given culture solves problems that the culture thinks are important. In an effort to shift those priorities, this article shares two examples, drawing attention to problems that we find both compelling and important: mercury poisoning and climate change. They can easily be used to enrich a single-variable calculus curriculum. The first involves tools such as modeling exponential decay or using geometric series, and grew out of a problem in [17] about a poisoned grain supply in the 1970s. The second applies numerical calculus to analyze data regarding a major methane leak in Southern California in 2015. For each of these examples, we provide background and context for the instructor to motivate students to tackle these two scenarios mathematically. We also provide, in two appendices, sample student handouts for instructors to use or adapt to their own contexts. Both sets of materials can be used for in-class examples, extended assignments, or group projects, depending on the time and student effort that can be devoted to them.

This article thus provides two concrete examples to deepen student understanding of the tools of calculus and, at the same time, to stimulate student engagement with the world around them. With some reworking either of the projects could be used in courses in differential equations or mathematical modeling. In our experience, when first using materials such as these in the mathematics classroom, instructors most benefit from assignments and projects that incorporate relatively self-contained motivation and background. This way they as well as their students can focus on the mathematics involved. In other words, students need not be doctors or lawyers or scientists to have a meaningful understanding of the issue at hand; the instructor is often provided with a more detailed background.

At the same time, students should readily see that there are connections to be made between the specific computations of their project and more global issues, such as pollution and climate change, which most threaten marginalized communities. Approaching these problems with mathematical tools, students will hopefully be better-equipped to confront similar challenges, especially in a world where corporate interests may not want to illuminate situations such as the ones presented here (witness the Volkswagen emissions scandal). We believe that students will thus be empowered not only by using mathematics to understand the scope of these situations but also by seeing mathematics as a tool for fighting for social justice. Our two books [11, 12] offer a wealth of course materials and other resources for college instructors who are seeking more ideas in this direction.

2. MERCURY POISONING: THE IRAQI GRAIN DISASTER

A safe food supply is essential to human well-being. Methylmercury is among the heavy metal toxins that have entered our food system. Students may be familiar with the issue of mercury in fish and perhaps have heard warnings to limit their weekly servings of tuna.[1] Because levels build up in the human body very slowly over time, the harmful effects of such toxins are often unnoticed, and these cautions may not be taken to heart.

Even those students familiar with, and careful about, the mercury accumulation problem in large fish will likely be far less familiar with the history of methylmercury in the food supply. A particularly devastating example occurred in Iraq in the early 1970s. In [17], this situation is presented in a word problem, using a geometric series to model the build-up of mercury and the levels at which devastating side-effects are experienced. With its compelling context, this problem is a notable outlier among the many geometric series problems found in textbooks. In what follows, we give background so that an instructor can motivate students to do more with this example. A sample assignment handout is provided in Appendix A. The interested reader can find further scientific detail in [1], an article from *Science*.

Methylmercury had been used as a fungicide for crops such as wheat, but as understanding grew of its toxic effects on human health, its use was banned in many countries beginning in the 1960s and 1970s. In 1971, some seed grain, that is, grain intended to be used as seed, as opposed to being fed to livestock or milled into flour to be later consumed by humans, was treated with methylmercurial fungicide, and was then sent to Iraq in late fall. This grain, in shipments of both wheat and barley, was distributed from the port of Basra across Iraq, its final destinations concentrated in the northern provinces such as Kirkuk. Meant only to be used for planting, this poisoned grain unfortunately entered the food supply, primarily through homemade bread, but also through meat from livestock and game who were fed the grain, contaminated soil, fish in water sources polluted by dumped grain, and even breast-feeding, where mothers had been exposed. A brief summary of this situation is provided in the handout in Appendix A, giving context for the problems which follow.

Students may wonder why this grain was used for food if it was not intended to be; a number of factors that may have exacerbated the epidemic are noted in [1]. Principal among these was the fact that symptoms of mercury poisoning do not manifest themselves quickly. In this case, seed seems to have arrived after farmers had already sown their fields, and so instead of waiting till the next season to plant the grain, some

[1]The primary sources of mercury in fish are industrial activities such as the burning of coal [6].

farmers tested it by feeding it to chicken. When there were no immediately visible ill effects, they assumed they could eat it too:

> ... some gave treated grain to their chickens for a period of a few days and observed no harmful effects. Human beings eating the contaminated bread may not have symptoms for weeks or months. By the time symptoms occurred, a toxic dose had been ingested [1, p. 239].

Meanwhile, warnings were not necessarily printed in the local language, and dye which had been used to mark the grain as seed grain could be simply washed off, which was taken to mean that the fungicide could be washed off.[2]

Although the human body can break down and otherwise eliminate some toxins, such as alcohol in the liver, mercury accumulates in the body over time. As mercury builds up, it has a profoundly toxic effect on the nervous system, causing loss of sensation (paresthesia), then loss of coordination and difficulty in walking (ataxia), slurred speech (dysarthria), hearing and vision defects, and even death.

As the number of casualties rose, doctors from the Medical College of Baghdad University were searching for treatments, and in an era long-predating e-mail and the internet, they wrote and mailed a letter to the *British Medical Journal* (now The BMJ), which was published there under the headline "Mercury Poisoning from Wheat" [7]:

> SIR — We would like to draw attention to an outbreak of poisoning from the mercurial compound Granosan M ... which has ravaged Iraq in the last two months .... The number of hospital-admitted cases exceeded 5,500 and the deaths reached 280 .... It will be greatly appreciated if doctors from other countries with experience in this field, particularly as regards treatment, would correspond with the undersigned.

By the following year, *Science* put the number of related hospital admissions at nearly 7000 and the death toll at over 450, counting only those deaths that occurred in a hospital. In some areas, as much as 10% of the population was hospitalized. Gradually over time, the body

[2]These various factors place this incident in a context of social justice, which is fundamentally focused on human well-being. For us, as in [12], "social justice" is a cornerstone of what makes a society good: individuals and communities living in such a society are all guaranteed to have certain fundamental rights. For the purposes of this article, the United Nations definition ("Social justice may be broadly understood as the fair and compassionate distribution of the fruits of economic growth" [15]) together with Thomas Jefferson's list of the unalienable rights of all people ("Life, Liberty, and the pursuit of Happiness") captures the essence of the term. The grain was never meant to enter the human food chain in the ways it ended up doing; there were factors which point toward responsible people not doing their job right (marking the grain permanently, using language and signage that would be intelligible to the local population, educating the farmers to the limited use conditions of the grain, etc.), thus disregarding the inalienable rights of the many people they harmed. What separates this from mere incompetence is the large extent of its impact on a whole community which did not have the power to protect itself.

excretes mercury, and some patients were treated with mercury-binding agents. Tragically, there was no treatment that consistently helped.

To better understand the progression of the poisoning, the build-up of methylmercury in the body can be modeled using a series. See [17, pages 499–500] for a concise approach, assuming a daily dose of bread (with 1.4 mg of mercury) and a daily excretion rate (0.09% of the quantity present). Using a geometric series, students can compute partial sums and deduce over how many days various symptoms may manifest themselves, and, in the long run, what levels of mercury are reached. In the table in Appendix A, we see the various thresholds at which symptoms develop, for situations ranging from numbness or even hearing loss to fatal levels of mercury. The actual daily dose varied in Iraq, depending on how much bread was consumed; the mercury level in the grain used (tested samples ranged from 3.7 to 14.9 μg of methylmercury per gram of wheat, with an estimated average of 1.4 mg per load); and differences in patient body weight. Depending on their familiarity with geometric series, students may or may not quickly recognize that they can model the total buildup of mercury in this way. The first two questions on the handout in the Appendix ask for the buildup after two days and then a week. It is typically helpful to have students construct a table of total mercury levels starting with one day, and continuing with the computations day by day, so that the geometric pattern becomes apparent. Indeed, instructors can start work on this handout with a discussion involving the entire class, inviting them to construct such a table, before dividing into smaller groups or assigning the questions in some other way.

In a footnote to the *Science* article, the authors model the amount of methylmercury accumulated, B, as a function of the amount of methylmercury ingested daily, m, an elimination constant, k, and time, t. Specifically,

$$B = \left(\frac{m}{k}\right)(1 - e^{-kt}).$$

This standard model is found in calculus texts typically to describe the amount of a drug in a patient's body over time. Assuming that at some point, the patient is no longer ingesting a daily dose of mercury, then the amount declines exponentially from its maximum value, where

$$B = B_{\max} e^{-kt}.$$

As is traditional for exponential decay, the elimination constant is related to the half-life, i.e., the time, $T_{1/2}$ for the mercury to decrease by half via

$$T_{1/2} = -\ln 2/k.$$

Based on a study of methylmercury using a tagged isotope, the half-life is estimated to be 76 days, although in the Iraq study it was found that this varied among individuals. This background and model is provided in the

Appendix handout so that students can compare this approach with their results using geometric series.

The mathematics above is certainly accessible to students in a standard calculus course. Furthermore, it is clear that mathematics can help us make some sense out of this situation. More specifically, in a calculus class, students can describe the human effects of the disaster using the mathematics they are learning. In this way, they can see the relevance and power of mathematics in contexts where human life may be at stake.

Appendix A provides further necessary data, sample questions, and a writing extension to contemporary issues. Lower threshold levels at which various symptoms may appear are provided, in terms of mg of methylmercury accumulated the body. The values given here, and used in [17], were extrapolated from levels of mercury in blood samples (see the study in Iraq [1]). In practice, such thresholds vary by patient; moreover, there can be a significant latency period before symptoms develop. As it would not be ethical to induce such toxic effects experimentally, it is difficult to find such data from other sources in the scientific literature, although the authors of the Iraq study noted concordance with estimates for paresthesia (30 mg compared with a range of 25–40 mg in Iraq) from a study in Scandinavia, based on consumption of fish.

The background to this situation is disturbing, and it should raise the stakes for the students. Rather than writing the scenario off as being in the far past and involving people far away, we can encourage our students to make bridges to current events. Connections can be made by students to modern problems such as mercury build-up from the consumption of fish and how information about health hazards is communicated or who has access to this information. Similarly, a connection could be made to the contaminated water supply in Flint, Michigan, as well as the danger of lead exposure faced by other communities. See Footnote 3 and the questions at the end of the next section for related ideas.

3. EMISSIONS AND CLIMATE CHANGE: THE ALISO CANYON METHANE LEAK

In late 2015, a severe methane leak began at a storage facility in the greater Los Angeles area. This was a situation for which there is *rate*

[3]This social justice issue is more precisely an environmental justice issue. The Environmental Protection Agency of the United States of America defines environmental justice as "the fair treatment and meaningful involvement of all people regardless of race, color, national origin, or income, with respect to the development, implementation, and enforcement of environmental laws, regulations, and policies" [9]. At the end of this section we propose some discussion questions instructors can use to bring up these dimensions of the issue explicitly in their classrooms.

data publicly available, which naturally lends itself to the tools of calculus. Although this topic is presented in the context of numerical calculus, students in a quantitative reasoning or general education class could also make use of the rate data by analyzing units and reason their way to an estimate for the amount of methane leaked. A sample assignment handout instructors may use or adapt to their own contexts is provided in Appendix B.

Below, we briefly provide background information for the instructor, culled from popular press coverage and *Science*. Articles from the *Los Angeles Times* ([2,14]) would make interesting companion reading for students, especially in explaining how the sample data were acquired.

The Southern California gas utility, SoCalGas, has a natural gas storage facility which uses old wells in Aliso Canyon, in the community of Porter Ranch, California. In October of 2015, the utility gave the State notification of a leak due to a failure in one of its storage wells. Although methane is not visible to the naked eye, for safety an odorant is added to it, and a powerful smell throughout the area clearly indicated the problem. Infrared scans, available on the Environmental Defense Fund website [8], illustrate the sizable gas plume above the well. The leak was ultimately described as massive – the largest in U.S. history and sizable relative to industrial activity as a source of methane emissions – and SoCalGas was required to pay short-term rental subsidies to Porter Ranch residents so that they could move out of the area from mid-November to the following February, when the leak was finally abated. Such a significant leak has negative impacts on health, air quality, and ultimately, the climate. Controversy continues in the community regarding SoCalGas clean-up efforts and the existence of the storage facility itself.[3]

In order to obtain an estimate of the size of the leak, a UC Davis scientist, Stephen Conley, flew back and forth through the plume at various heights, using specialized sensors and sample canisters to make methane measurements. From as many as 35 tightly-turned passes through the noxious air, measurements at different heights allowed Conley's team to estimate the rate of the leak at that time. Findings were published in *Science* [5]. Rate data computed for the dates of flight measurement and published via the California Air Resources Board forms the basis for student study. (See [3] for data values and [4] for an overview.)

A sample assignment handout is provided in Appendix B. Following our philosophy of directing the students immediately towards using mathematical tools, the problem has been stated with a minimum of background detail. This way students can immediately jump into the numerical calculus problem. The emphasis is on having students explain the approach they took. In practice, students usually quickly realize that they must take a numerical perspective and cannot simply compute an anti-derivative (despite all their practice doing so). Keeping careful track of units is often the main issue for students, and for this reason the problem of computing the mass of methane leaked would in fact also be well-suited to a quantitative reasoning course.

This simple numerical calculus problem easily lends itself to extensions: students can read related coverage from the *Los Angeles Times* (see [2] and [14]) and write a report on the situation, including their findings, for a lay person. Unfortunately, the impact of methane with respect to climate change is significantly worse than carbon dioxide. Students can research the impact of methane in comparison with carbon dioxide, and thus estimate the equivalence for emissions in terms of miles driven, for example, making standard assumptions about fleet averages for the United States. (The Environmental Defense Fund has similar figures presented at [8].) Alternatively, a comparison can be made to the impact of methane from the dairy or meat industry, for example, estimating the comparable number of cows. A few questions in this vein are posed in the sample handout and were used as an extra credit opportunity by one of the authors.

Instructors can also consider extensions for a class where students are asked to do more writing or to take on philosophical and ethical questions. For example: Porter Ranch is an affluent community, and there has been speculation about whether the utility would have responded the same way in other municipalities. What if methane did not have a pronounced odor? Would the community have been as alarmed? How responsive have authorities been to health hazards elsewhere, such as the lead in the water supply in Flint, Michigan? In particular do environmental hazards threatening marginalized communities get addressed as promptly or effectively as they were in the Porter Ranch community? Some followup readings such as [10] and [13] may help students and faculty address these issues and their connection to environmental justice more comprehensively.

4. STUDENT REACTIONS AND CLOSING THOUGHTS

When we used these and other similar materials in our classes, we have often received (unsolicited) student feedback that strongly indicates increased interest and curiosity. In particular, students working on these projects wanted to know how their results compared to those of scientists, even at times when they had not been explicitly asked to make such a comparison; they frequently referred back to these topics during discussions later in the term; and in multiple semesters, these topics (either the methane leak or the mercury-poisoned grain) were consistently chosen by a number of students during an unrelated assignment which required each individual student to choose any mathematical topic from the semester to discuss with a family member, peer, or friend who was not in the class.[4] These all demonstrate, at least on an anecdotal level, high levels of student engagement with this material. Indeed, students have sometimes shared the opinion that these were among the first examples they had seen where mathematics was focused on an issue that they view as a real problem in the real world. Such responses are consistent with those noted in [16], where students reflected on the study of environmental sustainability and income inequality in the context of a calculus class.

Even though we have not formally assessed the impact of this material in our classes, we are optimistic about the implications. Still we ask ourselves: Will these students now read the newspaper with a sharper eye, on the lookout for claims and supporting data they can analyze for themselves? At this point we are merely hopeful. In the future, it would be valuable to see if through these sorts of projects, instructors can accomplish the pedagogical goals not only of increasing their students' understanding of the tools of calculus but also of empowering students and supporting their inclination to apply those tools to the world around them.

ACKNOWLEDGEMENTS

This work was inspired by a joint project between the authors on Mathematics and Social Justice and by a problem of Dennis Zill (emeritus, Loyola Marymount Univerity). The second author thanks the organizers of the ACS Workshop on Mathematics and Social Justice, Zeynep Teymuroglu (Rollins College), Joanna Wares (The University of Richmond), Carl Yerger (Davidson College), and Catherine Buell (Fitchburg State University), where the material on the Aliso Canyon methane leak was first presented.

[4]The requirement to discuss a topic from class with an outsider was in fact part of an intervention to thwart stereotype threat. An implicit goal of the intervention was for students to experience a feeling of domain expertise which their audience may not have had and thus cultivate a stronger sense of belonging, in taking on the identity of a mathematician or scientist for themselves. It is notable that the topics with a social justice bent were the very ones which many students gravitated towards.

APPENDICES

APPENDIX A: SAMPLE STUDENT HANDOUT, MERCURY POISONING IN IRAQ

Methylmercury in food[5]

In the fall of 1971, large shipments of wheat and barley grain treated with a fungicide derived from mercury were distributed within Iraq. The grains were treated with a dye to ensure that the farmers using them would know they were toxic for direct human consumption. However, due to various local reasons, a large proportion of this grain was eventually consumed by the farmers who received them. This led to an epidemic of mercury poisoning. Some of the ailments associated with mercury ingestion in relation to the level of exposure can be seen in Table A1.

Table A1. Mercury poisoning: Lowest threshold in body at which symptoms may appear

Condition	Lowest threshold, mg of mercury
Numbness in fingers and toes (paresthesia)	25
Loss of coordination, difficulty in walking (ataxia)	55
Slurred speech (dysarthria)	90
Hearing loss	170
Death	200

Source: Bakir et al, [1, p. 238].

In this project you will explore the human implications of this disaster.

1. Assume that a daily serving of bread contains 1.4 mg of mercury and the human body processes, breaks down, and eliminates consumed mercury at a daily excretion rate of 0.09% of the quantity present. Compute the amount of mercury in a person exposed to contaminated bread for 2 days.
2. Compute the amount of mercury in a person exposed to contaminated bread for a week.
3. Come up with a formula that describes the amount of mercury in a person exposed to contaminated bread for n days assuming daily

[5]This project was inspired by a related project in Single Variable Calculus by Dennis Zill [17]; see pages 499–500 therein.

servings of bread as above. What happens as n goes to infinity? What does this mean?

4. Given the ailments associated with specific levels of exposure, when might a person consuming contaminated bread regularly show signs of each ailment listed in the table? Are all of the thresholds met?
5. If a person stops consuming contaminated bread the first day ataxia symptoms are observed, how many days does it take for the mercury levels in the person's body to go below 25 mg?
6. Mercury poisoning is very dangerous. Blood levels of more than 200 mg may lead to death. At the daily consumption rate above, when would death be a possibility? Iraqi grain deliveries began in September or October; by January, the government was issuing stronger warnings to avoid ingesting the grain. What would the daily consumption rate be for death to be a possibility with 3 months of exposure?

The numbers we used above were estimates. Let us now review some research. Suppose the amount of methylmercury accumulated, B, can also be modeled continuously, as a function of the amount of methylmercury ingested daily, m, in mg; an elimination constant, k; and time, t, in days. Specifically,

$$B = \left(\frac{m}{k}\right)(1 - e^{-kt}).$$

Assuming that, at some point, the patient is no longer ingesting a daily dose of mercury, then the amount declines exponentially from its maximum value, where

$$B = B_{\max}e^{-ks}$$

where s is in days since the patient has stopped ingesting mercury. In both models, the elimination constant k is related to the half-life, $T_{1/2}$, i.e., the time for the mercury to decrease by half:

$$T_{1/2} = -\ln 2/k.$$

Based on a study of methylmercury using a tagged isotope, this half-life was estimated to be 76 days, although in the Iraq study it was found that this varied among individuals.

7. How do these numbers and formulas change your answers to Questions 1–6 above ?

Remark: The actual daily dose varied in Iraq, depending on: how much bread was consumed, the mercury level in the grain used (tested samples ranged from 3.7 to 14.9 μg of methylmercury per gram of wheat), and differences in patient body weight. In some areas, as much as 10% of the population was hospitalized; by the end of this epidemic, over 450 people had died.

Assignment Extension

How do these ideas translate to our understanding of the EPA recommendations for fish consumption? Specifically, what recommendations can you find regarding the consumption of tuna, and can you relate them to computations for the build-up of methylmercury? Be sure to use reputable sources of information and to cite these sources.

Write a short summary with findings to inform those who are unaware of the hazards of mercury.

OR

Should school cafeterias post warnings about tuna and mercury consumption? Briefly argue for or against this, basing your argument on data and what you have learned about modeling the hazards of methylmercury.

APPENDIX B: SAMPLE STUDENT HANDOUT, METHANE LEAK

Group project: Porter Ranch area methane leak

In 2015, it was revealed that a major natural gas leak was happening at the Aliso Canyon natural gas storage facility. At the time, you may have seen or heard news about this, affecting the Porter Ranch area which is near Los Angeles, California. The leak was so profound that the local utility, SoCalGas, paid for temporary housing so that residents could move out of the area for many months.

Among other impacts, the leak resulted in a major release of methane into the atmosphere. According to EPA reference materials, methane is the second-most-prevalent greenhouse gas emitted in the United States from human activities, and has a 25-times greater impact (pound for pound, over a 100-year period) than carbon dioxide (CO_2) emissions on climate change.

In order to estimate the severity of the leak and the amount of methane emitted, periodic measurements were conducted using small planes with monitors, capturing an emission rate each day that a flight was made. It was only in early 2016 that the leak was finally plugged.

Using these actual measurements, as measured by Scientific Aviation and released to the public by the California Air Resources Board, your goal is to estimate the total number of kilograms of methane emitted. You may use either a rectangular or trapezoidal approach. Note, however, that the measurements were not taken at regularly spaced time intervals, so you cannot assume that Δt is the same throughout. Either you can take each measurement as representing the height of a rectangle along with an appropriate base, or you can take successive pairs of measurements as the heights of the left and right-hand sides of trapezoids. You may assume that the leak rate for the two weeks before the first measurement was also 44,000 kg/h and that the flights were taken at roughly the same time of day on each date.

This is not a "rigged" example – you are using real data obtained to study the problem.

Your write-up must include: a paragraph explaining the approach you took and why it was appropriate; the computations that you conducted, indicating especially how you set things up (you do not need to write out every data value, but it should be made clear to the reader how you used the data); and your final overall estimate, with units. Assume that you are a team of environmental consultants researching the situation for the community, but that this particular report includes the technical details behind how you made your calculations. In other words, this is a report aimed at someone who wants to know exactly how you used the tools of numerical calculus, not a layperson. Your work will be graded on clarity of method/explanation, not simply a final answer.

Before beginning your computations: Review this assignment and the *Los Angeles Times* background article and decide on a plan for solving the problem.

As mentioned above, you should assume that the leak rate for the two weeks before the first measurement below was also 44,000 kg/h.

Data from Scientific Aviation is listed in Table B2.

Table B2. Data collected by Scientific Aviation

Date of Flight day-month	Leak rate measured kg methane/h
7-Nov	44,000
10-Nov	50,000
28-Nov	58,000
4-Dec	43,000
12-Dec	36,000
23-Dec	30,300
8-Jan	23,400
12-Jan	21,500
21-Jan	19,600
26-Jan	20,700
4-Feb	20,600
11-Feb	950

Your computations should be based on the data values above. Below (Figure A1) is a graph of the data, but do not make estimates by eye from the graph alone. The line segments connecting the data values form a visual guide but do not represent actual measurements in between the flight dates.

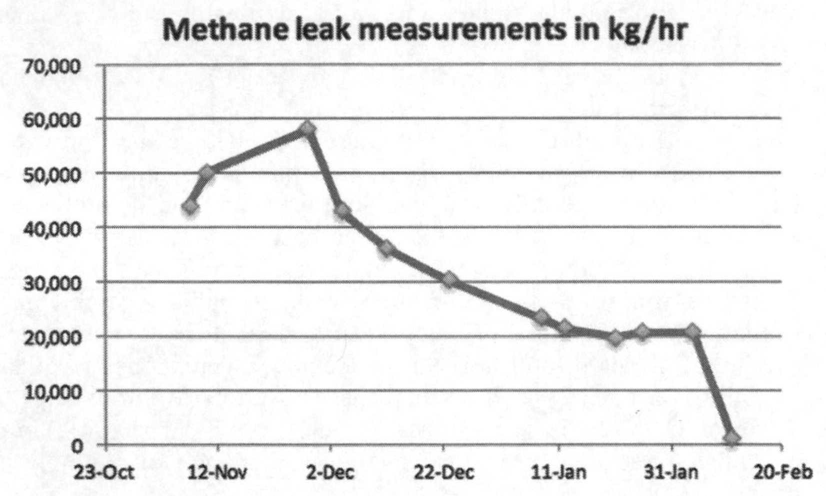

Figure A1. Leak of methane at Aliso Canyon facility; data available from the California Air Resources Board.

Assignment Extension

What is the impact of the leak? Find information about the potency of methane as a greenhouse gas to gauge this leak. Include your findings in a report, being sure to cite the sources of any background information. In particular, how does this leak compare to one of carbon dioxide? If these emissions had instead been the result of driving (in terms of carbon dioxide) or of the cattle industry (in terms of methane), what would the Aliso Canyon methane leak be comparable to?

Explain how your findings could translate to miles driven (or cars driven annually) or to number of cows raised for dairy or meat consumption. For the first conversion, you will need a comparison of the impact of methane versus that of CO_2, along with the average fuel economy for vehicles in the United States (and miles driven annually). For the second, you will need information about the typical methane production of cows, choosing either dairy cows or beef cattle. In either case, you may find different estimates for the information you need and may choose to give lower and upper estimates for the comparison. Be sure to use reputable sources of information for whatever comparisons you choose to make.

ORCID

Gizem Karaali http://orcid.org/0000-0002-0502-8358
Lily S. Khadjavi http://orcid.org/0000-0002-8808-3618

REFERENCES

1. Baker, F., S. F. Damluji, L. Amin-Zaki, M. Murtadha, A. Khalidi, N. Y. Al-Rawi, S. Tikriti, H. I. Dhahir, T. W. Clarkson, J. C. Smith, and R. A. Doherty. 1973. Methylmercury poisoning in Iraq. *Science*. 181(4096): 230–241.
2. Barboza, T. 2016, January 24. How much damage is the Porter Ranch leak doing to the climate? http://www.latimes.com/science/la-me-porter-ranch-greenhouse-20160124-story.html. Accessed 14 April 2017.
3. California Environmental Protection Agency Air Resources Board. 2016, February 11. Aliso Canyon natural gas leak preliminary estimate of greenhouse gas emissions (as of February 11, 2016). http://www.arb.ca.gov/research/aliso_canyon/aliso_canyon_natural_gas_leak_updates-sa_flights_thru_feb_11_2016.pdf. Accessed 14 April 2017.
4. California Environmental Protection Agency Air Resources Board. 2016, October 16. Aliso Canyon natural gas leak. http://www.arb.ca.gov/research/aliso_canyon_natural_gas_leak.htm. Accessed 14 April 2017.
5. Conley, S., G. Franco, I. Faloona, D. R. Blake, J. Peischl, and T. B. Ryerson. 2016. Methane emissions from the 2015 Aliso Canyon blowout in Los Angeles, CA. *Science*. 351(6279): 1317–1320.
6. Conniff, R. 2016, November 23. Tuna's declining mercury contamination linked to U.S. shift away from coal: Hard-won reductions in the environmental toxin could be erased if Trump proceeds with plans to resuscitate the coal industry and abandon climate initiatives. http://www.scientificamerican.com/article/tunas-declining-mercury-contamination-linked-to-u-s-shift-away-from-coal/. Accessed 14 April 2017.
7. Damluji, S. F. and S. Tikriti. 1972. Mercury poisoning from wheat. *British Medical Journal*. 1(5803): 804.
8. Environmental Defense Fund. Aliso Canyon leak sheds light on national problem https://www.edf.org/climate/aliso-canyon-leak-sheds-light-national-problem. Accessed 14 April 2017.
9. Environmental Protection Agency. Environmental justice. https://www.epa.gov/environmentaljustice. Accessed 16 February 2018.
10. Gray, W. B., R. J. Shadbegian, and A. Wolverton. 2010. Environmental justice: Do poor and minority populations face more hazards? https://www.epa.gov/sites/production/files/2014-12/documents/environmental_justice_do_poor_and_minority_populations_face_more_hazards.pdf. Accessed 16 February 2018.

11. Karaali, G. and L. S. Khadjavi, Mathematics for Social Justice: Perspectives and Resources for the College Classroom, MAA Classroom Resource Materials, American Mathematical Society, 2019.
12. Karaali, G., and L. S. Khadjavi, Mathematics for Social Justice: Focusing on Quantitative Reasoning and Statistics, MAA Classroom Resource Materials, American Mathematical Society, forthcoming.
13. Katz, C. 2012. People in poor neighborhoods breathe more hazardous particles. https://www.scientificamerican.com/article/people-poor-neighborhoods-breate-more-hazardous-particles/. Accessed 16 February 2018.
14. Khan, A. 2016, February 25. Porter Ranch leak declared largest methane leak in U.S. history. http://www.latimes.com/science/sciencenow/la-sci-sn-porter-ranch-methane-20160225-story.html. Accessed 14 April 2017.
15. United Nations International Forum for Social Development. 2006. Social justice in an open world: The role of the United Nations. http://www.un.org/esa/socdev/documents/ifsd/SocialJustice.pdf, accessed on 5 May 2017.
16. Verzoza, D. 2015. Reading the world with calculus. *PRIMUS*. 25(4): 349–368.
17. Zill, D. and W. S. Wright. 2011. *Single Variable Calculus: Early Transcendentals, Fourth Edition*. New York, NY: Jones and Bartlett Learning.

Supermarkets, Highways, and Natural Gas Production: Statistics and Social Justice

John Ross and Therese Shelton

Abstract: We present several modules that address social justice issues in an introductory statistics course. The activities consider possible disparities of housing location, language spoken at home, and job sector as they relate to, respectively, access to healthy foods, air pollution via proximity to traffic, and health concerns via proximity to fracking sites. Statistical content includes basic descriptive statistics, survey design and analysis, contingency tables, independence, various quantiles, and hypothesis testing. We describe the assignments, which satisfy some GAISE Guidelines, and we report on their implementation. Along the way, we share the challenges we faced and overcame as novices in the social justice community.

1. INTRODUCTION

The intersection of math and social justice is an important but often overlooked topic. In their survey of this intersection, Bond and Chernoff posit that this is, in part, due to:

> [t]raditional stigmas [which] have led many to view mathematics and social justice as being positioned on opposing ends of a spectrum describing quantitative and qualitative reasoning and, thus, unsuitable for integration. [5]

Although the article focuses on K-12 education, this viewpoint is present even at the college level, where social justice (SJ) is more commonly dealt with

Color versions of one or more of the figures in the article can be found online at www.tandfonline.com/upri.

in the Humanities and Social Sciences. Recently, we heard the claim that "Math is just about as far from social justice as you can get."

As mathematics professors and novice SJ advocates, we wanted to refute the standard narrative and introduce elements of SJ into our math classrooms. This paper documents our efforts, which centered on expanding our repertoire of statistical activities for students and improving the relevance of the content in our Introduction to Statistics classes, each of approximately 30–40 students.

As beginners in SJ, we were worried about issues such as managing our SJ classroom discussions or inadvertently reinforcing stereotypes. Our fears were assuaged by participating in a 2016 SJ workshop, where we found guidance and inspiration from like-minded SJ advocates. Due largely to their guidance, we overcame our fear of "going about it wrong," and designed a gentle introduction to statistical investigations of social justice – for the students, as well as for ourselves. We sought ways that statistics can answer the following question: Is there evidence of a disparity? Moreover, we used reasonably "safe" materials, ones that did not require difficult conversations and had less chance of being emotional triggers for students. The activities we created were of varied lengths and time commitments, allowing their implementation to be adaptable. All were largely successful. We improved our adherence to some of the national guidelines, specifically the Guidelines for Assessment and Instruction in Statistics Education (GAISE). We were able to: "teach statistical thinking, focus on conceptual understanding, integrate real data with a context and purpose, [and ...] foster active learning."[8] We adopted a guided inquiry-based learning approach to provide necessary scaffolding, and we broke activities into manageable parts. We say more about GAISE later.

In this paper, we outline the activities created, including implementation and grading. We discuss a Supermarkets assignment fully with more abbreviated discussion for another two. We conclude by discussing our students' reactions to these activities, as well as our own reactions. (See https://people.southwestern.edu/~shelton/MathSJFiles/indexMathSJFiles.) for the full assignments that we continue to improve. Our hope is that our narrative and our materials can provide a guide for other SJ novices, while our activities and collection of sources could be expanded for a more SJ-intensive experience.

1.1. Why intro stats and social justice?

Statistics is particularly well-suited for investigations of SJ for a number of reasons, including breadth of students affected and appropriateness of the field. SJ needs statistics, and SJ concerns are a great topic for statistics – or, more eloquently:

> The study of social justice is increasingly in need of empirical methods to describe, defend, and advise the critical analysis of the systems of domination and subjugation that permeate human power structures. The study of mathematics,

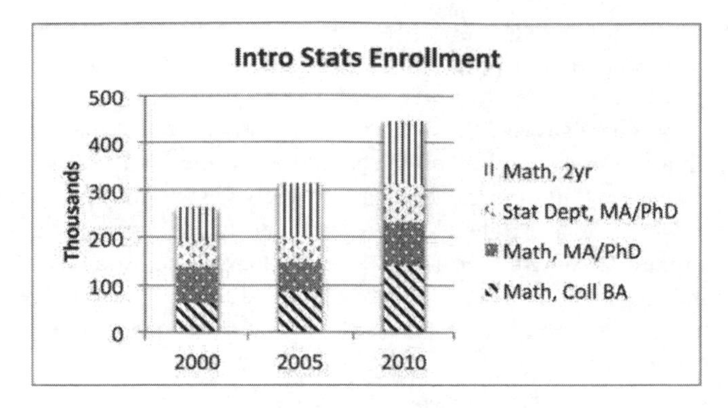

Figure 1. Enrollments in elementary statistics.

which often needs a meaningful context in which abstraction and anxiety can be nullified, is the ideal symbiotic partner for the study of social justice. [5]

The community of statistics educators has a large audience to reach, and that audience is growing. We present national data, derived from [3], [6], and [7]. From Figure 1, it can be seen that elementary statistics enrollments accounted for 10.2% of undergraduate mathematics or statistics enrollments in Fall 2010, for a one-semester total of over 446,000 students, including distance learning enrollments. In math or statistics departments, enrollments in introductory statistics grew 42.5% overall between Fall 2005 and Fall 2010, which includes 20.7% growth at 2-year colleges and 62.8% in math programs at 4-year institutions. Between Fall 2000 and Fall 2010, the overall increase was almost 71%. A linear trend of this data predicts a conservative 2015 enrollment of over 500,000 students.

The number of students taking the Statistics Advanced Placement (AP) exam has also increased dramatically, see Figure 2, though about 57% achieve

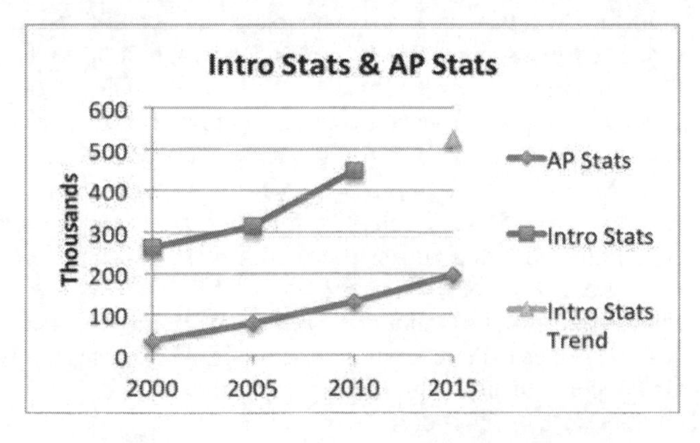

Figure 2. College enrollments and AP exam takers.

a score of three or better that is accepted by most colleges and universities for course credit, and 32% achieve the more selective scores of four or five.

These numbers suggest that introductory statistics enrollments are growing, and that the mathematical backgrounds of enrolled students are diverse. As instructors, we will often be teaching students for whom this is a terminal, "advanced" math class, right alongside students who have previously seen statistical material. Whether or not any of our students has some AP Statistics background, and whether or not Introduction to Statistics is the only math class a student takes, it must be our goal to provide as relevant and powerful an experience as possible.

> Students need socially critical mathematics and local, national, and global communities need students (i.e., citizens) who can interpret, articulate, and act upon social justice issues with the science of mathematics at their command. [5]

1.2. Our teaching environment

Our institution, Southwestern University, is an excellent environment to introduce these types of activities. Our small 4-year liberal arts and sciences institution lists SJ as one of five core values. Moreover, a 3-year institutional grant from Howard Hughes Medical Institute provided some support with funds for travel and module creation. Active inquiry methods of instruction are encouraged.

Our single flavor of elementary statistics aims to provide a foundation for students of all majors that require statistics, as well as for those who take it for general education; it cannot count toward a mathematics major. Seventy-five percent or more of our students in the statistics classes are in their first year of college. Through collaborative engagement, even those who take our course during their first semester will hear connections to SJ in other courses, and all students know they will take at least one course with a focus on SJ.

We are located in a suburban area surrounded by a mix of urban and rural communities. Our student body is primarily 18–22 years of age, though we are a veteran-friendly campus and have made efforts to be more open to transfer students. Most of our students are full-time, and most students live on or near campus.

Our Introduction to Statistics class is initially capped at 35 students. The relatively small class size and our institutional values most likely contributed to the success of our projects.

Our classes meet for 150 minutes a week, usually in two 75-minute sessions though sometimes in three 50-minute sessions. Although roughly 3 hours of contact is typical at many institutions, a longer time would be beneficial for in-class activities such as those described here.

1.3. GAISE

We include a bit more about the GAISE guidelines [8], especially for those who are not familiar with them. Statistics educators completed the first college guidelines in 2005, endorsed by the American Statistical Association. The Guidelines for Assessment and Instruction in Statistics Education for PreK-12 were formulated by 2007 in recognition of the inclusion of statistical concepts for some students at all levels. The college guidelines were revised in 2016. We focus on the college report, which includes many references on statistics education for the interested reader, in addition to learning goals, and sample activities and assessments. Included in the latest guidelines are the following recommendations.

1. Teach statistical thinking.
   - Teach statistics as an investigative process of problem-solving and decision-making.
   - Give students experience with multivariable thinking.
2. Focus on conceptual understanding.
3. Integrate real data with a context and purpose.
4. Foster active learning.
5. Use technology to explore concepts and analyze data.
6. Use assessments to improve and evaluate student learning.

For anyone relatively new to these recommendations, we point out that "multivariable" does not refer to multiple regression, but rather investigations of how data can be characterized by multiple attributes, and that different patterns may emerge when we consider the attributes. Students might experience this in comparing statistical graphs of data, grouped by geographic region, for two different years, for instance.

We recognize, as is pointed out in the GAISE document, that there are many versions of statistics that are currently taught at the college level. Our readers may emphasize particular methods, especially for courses that are more discipline-specific. We hold that "mastering specific techniques is not as important as understanding the statistical concepts and principles that underlie such techniques."[8] The activities we present here incorporate most of the recommendations, though more could be done (and was done in other assignments) with technology and assessments, in particular. We comment further after describing the activities.

2. THE SJ TOPICS

In this section we briefly outline the topics used to construct our activities. Each subject is centered around a professional report that touches on SJ issues. The activities ask students to read the report, answer qualitative questions about the importance of the topic of study, and answer quantitative questions about the given summary data. Here we discuss the SJ issues addressed and the statistical topics covered. Multiple strategies for implementation are addressed later.

2.1. Access to healthy food

Many students take healthy food for granted and do not think about food access as a SJ issue. Multiple reports have been published on food choice, food security, nutrition assistance, disability status, obesity, and more; see, for instance, the many online sources available from [1]. We created activities focused on survey analysis, quantiles, and conditional probabilities for two reports in which access to healthy, affordable food is proxied by distance to supermarkets. We chose to focus on the disparity in access between rural and urban areas, as this demographic split seemed less likely to be a sensitive topic. Activities designed for use with a long report [13] are discussed in this article. The rich document has 16 tables in the main article, and there is so much information that we will be able to continue to create other assignments beyond those shared here. In our activities, students grappled with surveys, various quantiles, and conditional probabilities. They slogged through a complicated table from the report with many variables to create a simple contingency table.

Another report [9] is just seven pages with three tables and includes information on education level. Students created and interpreted bar graphs to see differences in food access based on education level for one type of community but not for another. This is another example of statistical multivariable thinking, per the GAISE guidelines.

2.2. Proximity to major highways

Housing proximity to highways serves as a proxy for exposure to traffic-related air pollutants. A report explores who, demographically, lives close to a highway. The connection to social justice is clearly spelled out:

> The purposes of this report are to discuss and raise awareness of the characteristics of persons exposed to traffic-related air pollution and to prompt actions to reduce disparities. [4]

In particular, "(t)he greatest disparities were observed for race/ethnicity, nativity, and language spoken at home" [4].

The highways report is a very reasonable length (five pages with a single table). The report deals with multiple variables and comparative percentages, including a correlation coefficient that students interpreted. We guided students to create a contingency table and think about statistical independence of variables. Students computed expected counts, as they would in a chi-square test, though that was not covered in their course. We also incorporated hypothesis testing. For the most part, students found this to be fairly intuitive, as we focused on conceptual understanding, a GAISE guideline.

2.3. Proximity to natural gas wells

Hydraulic fracturing (also known as "fracking") is a method of extracting natural gas reserves in shale deposits thousands of feet underground. The extent of fracking has increased substantially in recent years, and there are growing concerns that fracking could contaminate local water supplies and pollute the air, ultimately resulting in negative public health effects (and other concerns, such as increased frequency and severity of earthquakes). We chose a study that assesses "the relationship between household proximity to natural gas wells and reported health symptoms" [10].

The study is very readable despite its length (19 pages before references and tables, double-spaced). Methodology used to conduct the sample survey is explicitly discussed, as are potential sources of bias. Therefore, this topic can be used as a launchpad for a class discussion of good and bad sampling procedures. The tables included at the end of the article compare demographics and various health conditions, grouped by distance to a natural gas well. Several are relevant to social justice: the report states a clear association between proximity to a natural gas well and an increase in health conditions. Statistical analysis can compare demographic information to location, such as the clear association between occupation class (blue collar, etc.) and proximity to a natural gas well. Descriptive analysis can add side-by-side graphs to the provided demographic information. Inferential analysis can include hypothesis testing or chi-squared tests to explicitly measure associations of proximity and health conditions.

3. IMPLEMENTATION

We designed several activities that examine the topics mentioned above for use in our Introduction to Statistics courses. These activities covered similar statistics and SJ topics in different ways and with different in-class time requirements, allowing the activities to be adaptable to different course styles and restrictions. One style uses in-class examination of the data, with students working in groups to answer questions. A second method employs a project format that did not take up any time in the classroom.

3.1. Food acquisition assignment with in-class portions

In three of our sections over two semesters, we began by implementing a multi-part assignment to complete individually outside of class which aimed to contextualize the article on healthy food access and provide experience with a small survey. The first portion consisted of Parts 1 to 3 below that produced survey data. The second portion consisted of data entry and class discussion, which are Parts 4 and 5 below. We describe the care taken to ensure that there was no risk of discomfort to any student. Parts 1 to 3 were completed outside of class.

Part 1: Students could complete an anonymous Food Acquisition Survey (FAS) that we designed from the report they would use, asking about household location (urban/rural), number of supermarkets in given proximity to the household, travel time to their most common food store, primary food source, and transportation to shop for food. We included the following rationale from the report and for the assignment [13]:

> Ensuring that Americans have adequate access to food is an important policy goal. In the 2008 Farm Bill, the U.S. Congress directed the U.S. Department of Agriculture (USDA) Economic Research Service (ERS) to learn more about food access limitations.

Additionally, our instructions included a motivational rationale:

> In order for us to better understand the survey process in general and the statistics involved, and to understand this study in particular, complete the following survey, or complete the alternative option.

Completing the survey anonymously was one option; another was to read text material about statistical bias, and then to critique whether our survey was biased. We provided an alternative to be sensitive to the feelings of even a single student, and we wished to make all students aware of the possibility of reactions by others. No one chose the second option, which, granted, would have entailed more work for the student.

Part 2: Students anonymously indicated the extent to which the survey made them feel uncomfortable, with the option to add an explanation. They could also turn in a blank sheet.

Part 3: In a non-anonymous portion, the student stated that she or he had completed the other portions and wrote a short paragraph about whether the survey in Part 1 could be used to answer questions taken from another report [14] about transportation options, distances traveled, type of store, and whether these differ by food security status (food security meaning "access by all people at all times to enough food for an active, healthy life"). This brief analysis leads students to think about survey design. The question about food security introduces and additional SJ issue without asking for such personal information.

The rationale in Part 1 and the questions in Part 3 provided a clear context and purpose, one of the GAISE guidelines. In our survey, we chose not to ask specifically about income bracket, food security, or participation in food assistance programs. We were aware that some of our students probably may have been uncomfortable answering such information, even anonymously. In a larger class, students might not feel quite as vulnerable, but in small or moderate-sized classes on a small campus, this could be an issue.

Part 3 was graded and returned. Grading was generous except for those who did not indicate they had done the other parts, or who did not read carefully and actually answered the questions instead. Participation points were given for students who indicated that they had completed other portions. Anyone who was late was given credit any time during the semester.

Part 4: When we taught two sections, survey responses from Part 1 and 2 were shuffled after removing any stray names. These were distributed to the students to process. Students entered the responses in a shared Google Sheet that had headers for the columns, as well as a sample entry. This provided students with an experience handling individual (rather than summarized) data, and in handling "dirty data." We purposely deflected their questions and had them decide how to enter unusual answers and write issues on the board. When we had just one section, the instructor entered the responses to better preserve anonymity and to ensure that no one had her or his own response, or the response of a friend.

Part 5: We facilitated a class discussion, beginning by listing issues students encountered in data entry. These included a response of a check mark instead of a single number; selecting both rural and urban; giving a range of values (three or four) instead of a single number; and no answer (is it zero?).

Students got to see firsthand about how little good data might actually remain from a survey.

Best of all, we discussed student responses to the survey reaction.

Instructions: Indicate the extent to which you agree with this statement: *"This survey made me feel uncomfortable."*

Results:

strongly agree, agree	neutral	disagree, strongly disagree	blank
5	14	76	1

The optional explanations were generally of the type indicating that grocery stores were not sensitive topics at all; some implied the question was ridiculous to ask. Some students, however, seemed aware of their position of privilege and were uncomfortable because of it. Very few said they did not wish to speak of such issues. Even a short discussion of these responses was very beneficial.

Additionally, we discussed the data in the survey (and later in the article) regarding number and type of variables, appropriate summary statistics, and good graphical displays. We created simple graphs during class. With our small class size, we felt single-variable analysis was appropriate, and that a larger class would be needed for multivariate analysis (even with making a two-way table) to preserve anonymity. Had we asked, "Has your household relied upon a food assistance program in the past year?" the comparison with discomfort would have been interesting, but insensitive.

A multi-semester collection of such surveys provides greater anonymity. In a truly small class or more intimate campus, portions of the assignment might not be feasible. The instructor might be able to connect with another institution to have a larger, pooled data set. Online forms could be used to provide greater anonymity and ease of data collection and entry. Resampling techniques could be used from such a data set. As an alternative, the class could read about the proportion or number of pre-college students who participate in food assistance programs, either at the state or national level, and discuss how students from different backgrounds might feel about survey questions.

3.2. Reading comprehension and guided data analysis

This assignment is also based on the report [13] regarding healthy food access. We offered it immediately after the FAS, with the hope that the FAS would help personalize and contextualize the guided data analysis; however, the assignments could be used independently from each other.

The activity asks students to read *portions* of the lengthy report. Students then answer a few basic questions and follow a structured guide that helps them process two of the tables from the report. Doing so should aid students who continue to struggle with the concepts of percentiles after covering quartiles. From Table 6 of the report, students interpreted *quintiles* of distance to a supermarket, based on other variables of vehicle availability and income category; some are shown in our Table 1.

Students are led to link the data to statements in the report, such as the following:

> Households without vehicles are closer to supermarkets in both low-income areas and in moderate- and high-income areas. Further, households in low-income areas are closer to supermarkets than households in moderate- and high-income areas, regardless of whether they have vehicles.

Table 1. Distance miles to nearest supermarket by vehicle access and income

Percentile	20	40
Low-income area		
No vehicle	0.04	0.28
Vehicle	0.26	0.57
Moderate/high-income area		
No vehicle	0.03	0.31
Vehicle	0.38	0.68

Table 2. Number of people (millions) with a moderate supermarket distance by region and income

	Low-income area	Moderate-/high-income area	Row totals
Urban area	28.5 (37.5%)	56.6 (37.9%)	85.1 (37.7%)
Rural area	1.8 (9.0%)	2.9 (4.7%)	4.7 (5.8%)
Column totals	30.3 (31.4%)	59.5 (28.3%)	89.8 (29.3%)
All areas	30.7 (31.9%)	62.4 (29.7%)	93.1 (30.4%)

Table 7 of the report concerns "Supermarket access for low-income areas compared with moderate/high-income areas, 2006 and 2010" and is wonderfully complex and full of information. Rows are the two stated income brackets plus a total for all areas, urban areas, and rural areas. There are three sections of columns: three percentiles (20, 50, 80) of distance to nearest supermarket for 2010; populations for both years; and a complex section – for each of three distance categories, and for each of two years, number of people in millions and the row percent.

For example, in our Table 2 we show information that students are guided to extract from the report's Table 7. This gives the number of people, in millions, and percentages of those who have a "moderate" distance to a supermarket in 2010. Students can figure out that the differences in the numbers for the last two rows are due to the distance categories: 10–20 miles for rural, but 0.5–1 mile for urban areas and "All areas." Percentages are provided in the report's Table 7, and each has a different denominator. For instance, 37.5% of the 76,000,000 people who live in urban, low income areas have a moderate distance to a supermarket whereas 37.9% of the 149,600,000 people who live in urban, moderate/high income areas have a moderate distance to a supermarket.

This assignment satisfies several GAISE recommendations, including providing students experience with multivariable thinking. Our questions focus on statistical literacy, but get students thinking about how to investigate possible disparities. Furthermore, students can apply the logic of joint and conditional probabilities when we ask these questions in the context of a situation; they rely on conceptual understanding, without formal notation or definitions.

In our implementation, a number of students needed help understanding the report's Table 7. However, some students felt an accomplishment in having processed this level of complexity with the aids of collaborations and of the scaffolding provided by the assignments.

3.3. Out-of-class projects

Another series of activities was created to introduce all three SJ topics through projects to be completed outside of class. For each project, students (working individually or in small groups) were asked to read the accompanying report, then answer a series of questions. The questions typically were split into several sections. The first section contained qualitative questions regarding the study detailed in the report. Students were asked to explain, using a short-answer format, the assumptions, reasoning, and methodologies of the original researchers. This led students to explore topics related to survey design and bias in the data collection. The second section was quantitative, and asked students to use data tables, present in the reports, to work with percentages and answer questions about independence of variables.

In the highway activity and the gas well activity, a third section was included that asked students to reflect on how the report and data were relevant to their personal experiences.

Students were given the option of whether to work individually or in groups. Projects were handed out at the end of one class period, and students typically had between 1 and 2 weeks to complete the project. Some students would come to office hours to discuss the project, but no time was devoted in-class to working on the project.

These projects were graded in a generous and straightforward manner. The short-answer problems from the first section were graded out of two points (where a grade of 1/2 would be issued if the student was on the right track but was missing a key concept of the problem, and a 0/2 would be issued if the student left the answer blank or grossly misunderstood the problem). The computational problems from the second section were graded out of one point (in the case of a short computation), or perhaps more points if the computation was more involved. Finally, the third section (which was a more reflective section) was graded on participation alone.

4. REACTIONS AND REFLECTIONS

4.1. Student engagement

The activities showed promising results for student engagement with SJ issues. Students found the reading and assignments challenging, but they were reassured by the collaborative nature of the work.

For the in-class activities, students worked in groups while the faculty member floated around the room to answer or ask questions of the various groups. We sometimes called the class together for a discussion about common issues with, or an insightful student remark about, the materials. We also called a meeting of a member from each group to compare answers and confer together, and then report back to the groups. These are examples of informal assessments. We used active, inquiry-oriented pedagogy to dynamically check and aid student understanding of statistics. This included guiding student comprehension of the statistics in the source reports.

The in-class time spent on this material was substantial. Our class periods were 75 minutes, and we used more than one class period (in some cases as many as 1.5 class periods). Students were also learning some new, and solidifying formerly learned, statistics material *through* the activities.

The out-of-class projects, of course, did not have the same in-class time commitment. The SJ elements of the reports were not discussed in class, so it was hard to measure student response to that aspect of the material. To attempt to measure this response, on the later projects we included a space for students to write their response to the report, with a prompt "Has this project (or this article) raised your interest in, or awareness of, the issues of [the topic in question]?" The problem was graded on participation only, so students could respond in any way they wanted (or at any length). This was made clear in the instructions, to encourage students to respond honestly. Several of the responses were very short (even one-word responses), and a small number answered negatively (that is, the articles did not increase their interest in the topics discussed). The majority of students responded positively, indicating that they had gained awareness of both the topics covered in the articles and the inequities and disparities examined in the projects. Several students made deeper connections to the projects, reflecting on their personal experiences and discussing how their upbringing and status informed their relationship with the issue.

Whether in-class or out-of-class, we found that most students seem to appreciate the relevance of the topics and were willing to engage with them honestly.

4.2. Faculty perspectives

We entered this experience with a level of reservation. Although we believed that teaching with an emphasis on SJ would improve our classes, we were concerned with our lack of formal training. It felt like many things could go wrong, and that we could do a disservice to our students with regards to these difficult topics. Ultimately, we found the experience to be incredibly positive, both for ourselves and our students. In particular, Introduction to Statistics seems a natural course for socially critical analysis, helping students to relate mathematics to the world around them.

Even before we committed to including SJ elements in our classes, we found a warm and welcoming SJ mathematics community. Sharing ideas and fears with like-minded individuals went a long way in preparing and motivating us. Undoubtedly, the encouragement we received in the SJ preparatory workshop was invaluable, including in sharing and calming our worries. We also found that working together was fruitful, giving us a small support structure at our institution.

We feel that the studies cited provide good "real world" data sets even for instructors not ready to try the more personal side of SJ. Introducing students to the statistical tools to evaluate social inequalities is important, and students can learn about the social implications even without asking for the students' own experiences, if one is hesitant to do that.

We were pleased with how our projects adhered to many of the GAISE guidelines. Following these guidelines, we "[used] assessments to improve and evaluate student learning" [8]. For each SJ topic, students submitted written work that we graded for statistical content. These formal assessments satisfied the GAISE criteria, "[g]ood items assess the development of statistical thinking and conceptual understanding, preferably using ... real data." In particular, the activities "[set] problems in realistic, meaningful contexts; they are data-based." We feel they could be improved in "[going] beyond calculation to probe deeper understanding of [statistical] concepts" and in using technology. Students were able to complete the worksheets with calculators. Other, non-SJ assignments, required statistical software.

5. THOUGHTS FOR FUTURE IMPLEMENTATION

Overall, we were pleased with our first attempts to include SJ issues in a mathematics classroom. Our activities, in either in-class or out-of-class form, were challenging to create, but felt relatively easy to implement. That being said, there are certainly places where we believe our projects and their implementation could be improved.

One major way that we can improve our implementation would be to more carefully measure how the projects improved or aided the students' understanding of the statistics topics covered, as well as students' learning/comfort with the statistical tools they were using. We observed that students participating in these activities learned the statistical content as well as in previous semesters. Additionally, we heard some comments that the real-world contexts made the statistics more interesting and meaningful, and the students' written responses seem to support that idea. In future implementations we hope to turn this anecdotal evidence into more concrete evidence, perhaps by means of a survey. In another, non-SJ assignment, we had students give feedback on their self-assessment of learning that we adapted from [12]. The following style could be used in the SJ assignments, also:

> For each question, on a scale of 1 for "Strongly Disagree" to 5 for "Strongly Agree", indicate your agreement. The activity added to my understanding of ... *(with a list of the statistical topics covered.)*

Either no grade would be attached, or any response at all would count as a participation grade. This would best be done for each assignment, but could be done at the end of the semester.

We also administered a mini "social justice pre-quiz" inspired by [11]. A pre-post comparison (facilitated by administering a similar quiz at the end of the semester) would have been beneficial, allowing us to more rigorously measure increases in SJ awareness made over the course of the semester.

We would also like to improve our implementation by expanding the scope of our projects. For example, we might use the portions of the Supermarket report dealing with travel areas, or analyze census data to compare income distribution by rural and urban areas. "(N)ationally, rural Americans have lower median household incomes than urban households, but people living in rural areas have lower poverty rates than their urban counterparts." Despite that, "American Indian and Alaska Natives living in rural areas had higher poverty rates than their counterparts in urban areas" [2]. Of course, we would be cautious focusing on any segment of our society related to income status or education level without a lot of preparation. We are afraid of further marginalizing an already marginalized group and of reinforcing stereotypes. We could collaborate or confer, at least outside of class, with a faculty member experienced with SJ discussions.

6. CONCLUSION

We believe that the SJ projects added a valuable context for our teaching and the students' learning of statistics. We feel emboldened by the success of this excursion, and hope this serves as a success story for others who are interested in, but nervous about, including social justice topics in their classes.

Acknowledgements

Many of the ideas presented here originated at an Associated Colleges of the South workshop, *Mathematics for Social Justice*, which took place at Rollins College, Winter Park, FL in May 2016. We were eager to attend this workshop, which prepared us to pursue the creation of these assignments. Most importantly, the workshop introduced us to a real community of mathematicians who care deeply about the issues of SJ within and beyond mathematics/statistics classrooms. Moreover, this community welcomes and supports engagement by any interested faculty, from those who are timid novices at social justice, as we were, through energetic and seasoned faculty. We gratefully acknowledge the organizers (C. Buell, Z. Teymuroglu, J. Wares, C. Yerger) and participants of this conference.

ORCID

Therese Shelton http://orcid.org/0000-0003-4909-0767

REFERENCES

1. Amber Waves. U.S. Department of Agriculture, Economic Research Service. https://www.ers.usda.gov/amber-waves.aspx. Accessed 22 May 2016.
2. Bishaw, A., and K. G. Posey. 2016. A comparison of rural and urban america: household income and poverty. http://blogs.census.gov/2016/12/08/a-comparison-of-rural-and-urban-america-household-income-and-poverty/. Accessed 3 February 2017.
3. Blair, R., E. E. Kirkman, and J. W. Maxwell., 2013. *Statistical Abstract of Undergraduate Programs in the Mathematical Sciences in the United States: Fall 2010 CBMS Survey*. Providence, RI: American Mathematical Society.
4. Boehmer, T. K., S. L. Foster, J. R. Henry, E. L. Woghiren-Akinnifesi, and F. Y. Yip. 2013. Residential proximity to major highways-United states, 2010. *Morbidity and Mortality Weekly Report, Nov 22, 2013. Centers for Disease Control and Prevention*. 62(03): 46–50. https://www.cdc.gov/mmwr/preview/mmwrhtml/su6203a8.htm. Accessed 1 June 2016.
5. Bond, G., and E. Chernoff. 2015. Mathematics and social justice: A symbiotic pedagogy. *Journal of Urban Mathematics Education*. 8(1): 24–30. http://ed-osprey.gsu.edu/ojs/index.php/JUME/article/viewFile/256/170. Accessed 13 February 2017.
6. College Board. 2010. AP program participation and performance data. https://research.collegeboard.org/programs/ap/data/archived/ap-2015, https://research.collegeboard.org/programs/ap/data/archived/2010. Accessed 6 April 2017.
7. College Board. 2014. AP report to the nation. http://media.collegeboard.com/digitalServices/pdf/ap/rtn/10th-annual/10th-annual-ap-report-to-the-nation-single-page.pdf. Accessed 20 December 2017.
8. GAISE College Report ASA Revision Committee. 2016. Guidelines for Assessment and instruction in statistics education. http://www.amstat.org/asa/files/pdfs/GAISE/GaiseCollege_Full.pdf. Accessed 9 February 2017.
9. Grimm, K. A., L. V. Moore, and K. S. Scanlon. 2013. Access to healthier food retailers - United states, 2011. *Centers for Disease Control and Prevention Supplements*. 62(03): 20–26. https://www.cdc.gov/mmwr/preview/mmwrhtml/su6203a4.htm. Accessed 23 May 2016.
10. Rabinowitz, P., I. Slizovskiy, V. Lamers, S. J. Trufan, T. R. Holford, J. D. Dziura, P. N. Peduzzi, M. J. Kane, J. S. Reif, T. R. Weiss, M. H. Stoe. 2015. Proximity to natural gas wells and reported health status: Results of a household survey in Washington County, Pennsylvania. Environ health perspectives. *National Institute of Environmental Health Services*. 123: 21–26. Accessed 1 June 2016.

11. Schmitt, B., and S. Quigley. 30 January 2012. A social justice quiz. www.counterpunch.org. Accessed 9 August 2016.
12. Spayd, K., and J. Puckett. 2016. A three-Fold approach to the heat equation: Data, modeling, numerics. *PRIMUS*. 26(10): 938–951.
13. Ver Ploeg, M., V. Breneman, P. Dutko, R. Williams, S. Snyder, C. Dicken, and P. Kaufman. 2012. *Access to Affordable and Nutritious Food: Updated Estimates of Distance to Supermarkets Using 2010 Data. Economic Research Report No. (ERR-143).* Washigton DC: U.S. Department of Agriculture, Economic Research Service. https://www.ers.usda.gov/publications/pub-details/?pubid=45035. Accessed 1 June 2016.
14. Ver Ploeg, M., L. Mancino, J. Todd, D. Clay, and B. Scharadin., 2015. *Where Do Americans Usually Shop for Food and How Do They Travel To Get There? Initial Findings from the National Household Food Acquisition and Purchase Survey. Economic Information Bulletin No. (EIB-138).* Washigton DC: U.S. Department of Agriculture, Economic Research Service. https://www.ers.usda.gov/publications/pubdetails/?pubid=79791. Accessed 1 June 2016.

Mass Incarceration and Eviction Applications in Calculus: A First-Timer Approach

Kathy Hoke, Lauren Keough, and Joanna Wares

Abstract: In the calculus classroom, integrating applications with the theoretical and procedural conveys the relevance of the material. Since Newton and Leibniz developed calculus in the context of physics and astronomy, applications in these fields tend to predominate. To engage and excite our students, pedagogical considerations must include broader examples from across the curriculum. Our students have been particularly responsive to examples that address social justice issues. Here we describe two projects used in calculus courses at two different (liberal arts) institutions concerning mass incarceration and eviction. In addition, we offer advice on the implementation for your classroom and include some of our materials.

1. INTRODUCTION

Newton and Leibniz formalized the mathematical subject of calculus in the 17th century while working on questions concerning planetary motion and other physical phenomena. Using calculus, Newton wrote *Principia,* in which he rigorously developed his laws of physical motion. For centuries following this, calculus was further developed and deeply integrated into the subjects of physics and astronomy. Due to these deep historical connections, many calculus classes still primarily use applications from physics to motivate calculus ideas.

In modern times, many of us have incorporated applications from other subjects into our calculus classes, as we have noticed that our classrooms have changed from predominantly science majors to majors from across the institution. To broaden our students' perspectives, we teach students that we can readily answer questions from chemistry, biology, and economics using material from beginning calculus courses. Students' interests in these subjects, as well as calculus itself, are piqued when we present the material side by side in this way.

However, there is evidence that looking for applications beyond the sciences, particularly applications related to social justice, benefits students. The three authors of this work recently attended a workshop called "Mathematics for Social Justice" at Rollins College (2016) where we began to learn about teaching mathematics in socially and culturally relevant ways. Just as applications in the sciences can pique students' interest in calculus, social justice applications can increase student motivation and buy-in. (See [11] and [24].) However, social justice applications can do more than pique interest in calculus. Incorporating social justice into the calculus classroom can help accomplish the following four goals (elaborated below): (i) to change students' views on mathematics; (ii) to equip our students with the quantitative abilities necessary to address social issues; (iii) to deepen our students' understanding of mathematics; and (iv) to create more inclusive STEM classrooms.

Among the benefits for students, of incorporating social justice into mathematics, is that they

> can recognize the power of mathematics as an essential analytical tool to understand and potentially change the world, rather than merely regarding math as a collection of disconnected rules to be rotely memorized and regurgitated [10, p. 2].

Thus, teaching with social justice topics is a way to combat the long-standing issue of students viewing math as a procedural subject with no creativity or real-world applications. Additionally, students' abilities to address social justice issues are elevated when they learn to provide a mathematical foundation for their arguments [9, 10]. Also, people have long argued that one of the purposes of education is to help our students become active participants in democracy [7]. Importantly for our mathematics courses, the pedagogical use of social justice applications does not "water down" the mathematics. In fact, students' understanding of fundamental mathematical concepts is deepened when the mathematics is framed within the context of something as interesting and current as social justice [10]. Finally, best practices for building inclusive and diverse STEM classrooms include recommendations for community engagement and altruistic motivation [6, 14].

In the semesters after learning about the importance of including social justice topics as applications in our classrooms, we made first attempts at incorporating materials with social justice examples into our calculus materials. Below, we delineate some of the common themes that appeared as we began exploring social justice applications, regardless of the social justice issue. We then give two examples of projects that intertwine calculus and social justice, developed and used in our calculus classrooms.

2. GENERAL APPROACH

Our exploratory approach to integrating social justice examples into our calculus classrooms has been to start with the mathematics of interest and then to look for questions of social justice that fit nicely with that mathematics. Since calculus is primarily about rates of change, we sought data that were changing with respect to some independent variable. The two projects in this article discuss incarceration and eviction rates (each changing with respect to the independent variable time). Data sets in this form are useful for teaching both first-semester calculus as well as second-semester calculus topics.

We chose our topics because they were contemporary and because we found appropriate data for these topics. When looking for social justice ideas and data, we found the following calculus topics particularly relevant: linear versus exponential growth, average rates of change, approximating instantaneous rates of change from discrete data, tangent line approximations, relationships between the graphs of a function and its derivative, second derivatives, and numerical integration. Despite discussing two different social justice issues in the same course, many of the questions we asked the students to discuss were the same.

Other common threads appeared in highlighting social justice issues in calculus:

1. *The use of technology.* Given that we were often working with discrete data (e.g., data by year), there was no function formula with which to take the derivative. This was a bonus, because it emphasizes that calculus is about rates of change and not just derivative rules. However, this also meant that it was helpful to use spreadsheet software, such as Excel or Google Sheets, to be able to compute, for example, many rates of change at once.
2. *Using primary sources.* Another common thread was the use of primary sources. Using primary sources enforces the reality of the topic. Moreover, if you can find a source that has drawn conclusions from the very data you are using, students can use this as a starting point: does the calculus we know reinforce or contradict the conclusions drawn in the source?

3. *Importance of units.* Students' ability to interpret the data seemed to depend heavily on their ability to state the units of the derivative or the approximation given by the tangent line. Encouragingly, we found that across the board, units became more interesting and important when discussing the following examples than we had previously seen using more classic calculus examples.

3. EVICTION AND RATES OF CHANGE

In 2016–17, the University of Richmond chose as its "all campus" read the book *Evicted: Poverty and Profit in the American City* by Matthew Desmond [8]. Campus-read programs are prevalent on campuses and this book naturally provided a theme to incorporate social justice ideas. Being part of a university-wide program also provided resources for framing discussions surrounding the book. Reading the book was optional for the calculus class, but extra credit was offered to students who read the book and participated in a small-group discussion about it at the end of the semester.

Although this book gives a very insightful and complete look at the people affected by eviction, its lack of data was a problem for bringing it into the calculus classroom. A report [20] produced by the San Francisco Anti-displacement Coalition entitled "San Francisco's Eviction Crisis—2015" provided an interesting, data-driven analysis of recent eviction practices in San Francisco. There were several similarities between this report and eviction practices in Milwaukee, WI, described in *Evicted.* Although not all of the report produced by the San Francisco Anti-displacement Coalition was used, it was required reading.

Issues surrounding eviction were incorporated into the first-semester calculus class via three data assignments, drawing from data referenced in this report, to accompany each of three tests. The first reinforced the ideas of linear versus exponential growth and average rate of change. The second reinforced the concept that the derivative is a rate of change and what we can learn about a quantity by knowing about its rate of change. Finally, the third project emphasized what we can learn from area under a derivative curve and numerical integration. The numerical integration part of third project was expanded and used in a second-semester calculus class as well.

The three projects used two data sets. The first data set consisted of median rents in the San Francisco area and percentages of renters in the San Francisco area that pay more than 35% of their income in rent. This data comes from the American Community Survey, an ongoing statistical survey by the US Census Bureau [23]. The second data set was the raw data used in [20]; it is data on eviction rates under the Ellis Act collected by the San Francisco Rent Board [19]. The Ellis Act is a state law that allows landlords to evict all tenants from a building if the landlord intends to stop renting the building and repurpose its use. The report [20] makes a data-driven argument (via a graph) that a steady increase in Ellis Act evictions was halted by the city's threat of a new tax on real estate speculation. When the proposition for the tax was defeated, Ellis Act evictions spiked. This data and this section of the report were chosen for two reasons: (i) the students could critique and mimic the report's graphical analysis; and (ii) recent articles in the *San Francisco Chronicle* [17], the *Wall Street Journal* [16], and CNN [13] relating Airbnb-like practices and rising eviction rates discuss effects of taxes or other regulations on the problem.

4. AN APPLICATION WITH MASS INCARCERATION

Another important social justice issue of our time is mass incarceration. This issue has been addressed in documentaries such as *The House I Live In* [2] and the recent documentary from Netflix *13th* [1], as well as books such as *The New Jim Crow* [3] and *Locked In: The True Causes of Mass Incarceration and How to Achieve Real Reform* [18]. Several politicians (including Rand Paul and Hillary Clinton) recently claimed that the US has less than 5% of the world's population, but 25% of the world's total prison population. (This has been verified, for example, in one article from the Washington Post [15].) In addition, according to conclusions from The Sentencing Project [21], there are large racial disparities in incarceration rates, and incarceration leads to a variety of other issues, including voter disenfranchisement.

This project was used in a 50-minute class (with students completing up to an additional hour of work outside of class). The students worked in groups of two or three. Conveniently, one can download data as a spreadsheet from the Bureau of Justice Statistics [5] that includes the total population under the supervision of adult correctional systems, by year, from 1980 to 2010. Unfortunately, this data does not show the stark differences in rates of change from before 1980 to after 1980. A more complete data set can be found as a PDF through the Sourcebook of Criminal Justice Statistics from University at Albany [22].

The assignment asks students to approximate derivatives (in Excel) using backwards, forwards, and central difference estimates, and to use tangent lines to predict future values. This material fits nicely with Section 1.5 of the free text *Active Calculus* [4]. Students were also asked to compare their tangent line estimate to actual data and explain any reason for disparity, hinting at the fact that not all growth is linear and how predictions are affected by increasing and decreasing rates of change. Thus, this assignment can lead into a nice discussion of the second derivative as well.

Before this lesson, the students had seen the following mathematical concepts, in other contexts: how to compute and interpret average rates of change, how the tangent line approximates function values, and the definitions of forward, backward, and central difference estimates. The content objectives for this lesson were to improve students' understanding of each of these ideas, as well as to introduce some context for the second derivative.

Many of the outcomes discussed in the literature cited in our introduction did occur. Although we generally found it rare for students to make connections to other classes, one student exclaimed "we talked about this in my sociology class!" On the final exam there was a question similar to those studied on the lab. The median score on this question was 10 of 13 (which was higher than the median on the final exam), suggesting that students successfully learned the content intended to be covered with the lab (and possibly better than other material in the class). Since social justice issues can be more charged than other applications, the social justice content was included as an open-ended bonus question in this first implementation. This helped avoid the potential pitfall of students reacting poorly to the inclusion of social justice issues in calculus.

5. REFLECTIONS

A question that many skeptics may ask is, why do this? Why not just teach calculus with traditional examples? We found that many of the potential benefits from the literature we cite in the introduction did occur in our classrooms.

One thing we discovered is that there were many students in our calculus classes who became more excited about the mathematics once they saw a purpose that went beyond basic business or physics applications. They liked the idea of learning a new way to think about calculus: in the context of the problems of affordable housing, unfair evictions, mass incarceration, or income inequality. These applications were far more motivating than any applications we had used previously.

In addition, emphasizing the broad range of applications of calculus keeps the mathematics relevant. Emphasizing the importance of interpretations shows students that Wolfram Alpha cannot do everything that we teach them, and that, sometimes, deeper analysis is required. The machinery of calculus needs a human brain to ask the right questions and use the right analysis techniques to answer them.

As teachers in liberal arts settings, we recognize the importance of educating the whole person. Exposing our students to how mathematics can, at its best, be used to make better policy and, at its worst, be used to confuse unsuspecting people, prepares our students for life, post-graduation, in a complicated world.

We have also received many questions on how to get started. One piece of advice is that it is okay to start small. We each began with identifying one concept that we could relate to a current issue. You can also start small within the issue itself. One option is to present the data for students to explore without having a whole-class discussion. One could also team up with faculty who have more experience with the social justice issue, but perhaps less experience with mathematics.

It is also the case that it is easier to get started in a calculus classroom that is conceptually based and already encourages students to critique answers, question assumptions, and justify reasoning [10]. These dispositions that we already encourage in our classroom make addressing social justice issues a more natural extension. In addition, since we ask our students to write and justify reasoning frequently, we were able to grade these assignments in a similar fashion to other assignments in the course. In particular, assignments were graded on mathematical correctness and reasoning.

Finally, one may feel that one needs to (at least be prepared to) defend the use of social justice issues and we felt that way as well. One convincing argument against this is that mathematics is already political [12]. In addition, Gutstein and Peterson make the argument that this is not something we should feel defensive about at all – "what we're talking about here is something that helps students learn rich mathematics, motivates them, and is really what math is all about" [10, p.5]. Most important for us while we were each trying new things in our classrooms, was having each other to offer support in the form of encouragement and constructive criticism.

ACKNOWLEDGEMENTS

The authors would like to thank the coordinators (Catherine Buell, Zeynep Teymuroglu, Carl Yerger, and Joanna Wares) and speakers (Nathan Alexander, Joseph Bolz, Lily Khadjavi, and Bonnie Shulman) from the ACS Social Justice Workshop that allowed us to meet and motivated us to get started. Many thanks to all participants in the workshop for the helpful and engaging conversations and to the Associated Colleges of the South Grants Program for funding the workshop. Last, but not least, we would like to thank the referees for the many helpful comments that greatly improved this manuscript.

APPENDICES

We include here sample assignments to be used in calculus classrooms. All associated files are available upon request.

APPENDIX A: SAMPLE ASSIGNMENT FOR CALCULUS ABOUT EVICTION DATA

1. Look at the first data set. These data give the percentage of renters in San Francisco that pay more than 35% of their income in rent by year. Let $H(t)$ be this percentage, where t is years since 2005. Thus $H(0)$ gives the percentage of renters that pay more than 35% of their income in rent in the year 2005.

   a. Make a scatterplot of $H(t)$ vs t in Excel or Google Sheets.
   b. Find the best fit cubic polynomial.
   c. Find the inflection point of your polynomial algebraically. (Show work.)
   d. Using the cubic polynomial model for $H(t)$, for what value of t is $H(t)$ increasing at the fastest rate?

2. Continuing with the same data as in #1, in this exercise we will use a DIFFERENT method to estimate the value of t at which $H(t)$ is increasing at the fast rate (1d). You will approximate $H'(t)$ using the data, using the closest AROC method. Then you will find the maximum value of $H'(t)$.

   a. Numerically estimate $H'(t)$ using "closest AROC." (See HW 19, #8.) Create a column in Excel with its values. Let t be years since 2005. Make a scatterplot with smoothed lines of $H'(t)$ in Excel or Google. (In Google, it works best if the column for t is next to the column for $H'(t)$.)
   b. What are the units of $H'(t)$?
   c. Use your scatterplot of $H'(t)$ to estimate the value of t at which $H(t)$ is increasing at the fastest rate. Explain how you found this value of t using the smoothed, marked scatterplot of $H'(t)$.
   d. How does your answer in c) compare with your answer in 1d)?

3. Returning to the report from the San Francisco Rent Board, let $e(t)$ be the total number of Ellis Act evictions that have occurred between April 1, 2010, and time t. The function we see graphed in the report (reproduced on next page) is a graph of the DERIVATIVE of $e(t)$ based on discrete data points. We can see that this graph is a derivative of something because its units (evictions per year) indicate that it is a RATE graph. In the second graph on next page, I have sketched a model for $e'(t)$ based on the first graph.

    a. Use the second graph (sketched model based on first graph) to describe what would look like by answering the following questions. When is $e(t)$ increasing? Decreasing? Concave up? Concave down? What are its inflection points? REMEMBER: you are looking at a graph of the DERIVATIVE of $e(t)$, not $e(t)$.

        a. For what values of t (approximately) is increasing?
        b. For what values of t (approximately) is decreasing?
        c. For what values of t (approximately) is concave up?
        d. For what values of t (approximately) is concave down?
        e. Find the approximate time(s) at which $e(t)$ has inflection point(s).

    b. What do you expect happened last year (2015–2016) to Ellis Act eviction rates in San Francisco? Why? What does the report from the San Francisco Rent Board predict?
    c. Open the second data set. This is the raw data that produced this graph, including updates for 2015–2016. Make a scatterplot of this data to see if your expectations in b) were correct. Read [16] and [17] for a possible explanation. Summarize this in a few sentences.

APPENDIX B: SAMPLE ASSIGNMENT FOR CALCULUS I ABOUT INCARCERATION DATA

Note: This activity is adapted from a similar activity concerning population written by colleagues at Grand Valley State University.

Part 1: Collecting and understanding the data

We will be studying the correctional population trends in the US.

One can find much of the data at http://www.bjs.gov/index.cfm?iid=2237&ty=pbdetail. You may download a spreadsheet of this data from Blackboard now (spreadsheet available upon request). Take a look at the data and answer the questions below.

This correctional population data is really a function that we're interested in. Fill in the blanks below.

The function _____ represents the ______ (measured in ______), ______ _____ after ____.

Part 2: Average Rates of Change

1. Compute the average rate of change of the number of prisoners between 1925 and 1980. Write a sentence that explains the meaning of the number you found including units.
2. Compute the average rate of change of the number of prisoners between 1980 and 2012. Write a sentence that explains the meaning of the number you found including units.
3. Are these rates of change consistent with population growth in the US? Why or why not? (You will need to look up more data to answer this question.)

   - For one extra point: do some research about the prison population in the US to find out what experts say accounts for the difference in rates of change. Cite sources.

Part 3: Wait! Calculate derivatives (together)

In this part, you will use your spreadsheet to calculate derivatives from your table of data. *We will use spreadsheet formulas in this part. Do **not** enter a lot of numbers by hand!*

1. In Column C, calculate the **forward difference estimates** for $P'(t)$. We will do this together. Use the space below to take notes about the formula. Do not enter the numbers directly. Instead, type the cell name (or click on the cell) to refer to its value. Hit return to end your input, and make sure the number is sensible.
2. Copy your formula from C2 down the rest of the column (as we have done before in Lab 1). You should see a list of forward difference estimates. Make sure they are sensible, and if not, debug your input.
3. Repeat with Column D, but calculate the backward difference estimates for $P'(t)$.
4. Average columns C and D to calculate the central difference estimates for $P'(t)$.

Part 4: Predict the future

Next, we will use the derivative calculations in Part 3 to predict future population trends.

1. Use the central difference estimates for $P'(t)$ for this question. When was $P'(t)$ largest? Closest to zero? Most negative? For each, give a year and then write one complete sentence to describe what this value says about the correctional population in that year. Be sure to include units.
2. Your spreadsheet should not have a central difference estimate for 2012. Write the formula for the line whose slope is the central difference estimate for 2011 and which goes through the point corresponding to the year 2011. (This is known as the tangent line.) Write this formula in your Word file in point-slope form.
3. The tangent line from (2) can be seen as an approximation of the function near the year 2011. Using your tangent line, estimate the correctional population in 2012. In your Word file, type the exact formula you used to do this. In 1 or 2 complete sentences, compare the value to the actual correctional population in 2012. Is your estimate an overestimate or an underestimate? What does this say about the rate of change of the rate of change?
4. Use your tangent line to estimate US correctional population population in 2020. Type the exact formula you used to do this (using Equation Editor!). In 1 - 3 complete sentences discuss how accurate you think this estimate will be (and why).

REFERENCES

1. *13th. Film.* Directed by Ava Duverney. Produced by Kandoo Films. Netflix, 2016.
2. *The House I Live In.* Film. Directed by Eugene Jarecki. Abramorama, 2012.
3. Alexander, M. 2012. *The New Jim Crow: Mass Incarceration in the Age of Colorblindness*. New York, NY: The New Press.
4. Boelkins, A., D. Austin, and S. Schlicker. 2014. Active calculus. https://activecalculus.org/single/. Accessed 31 December 2018.
5. Bureau of Justice. Statistics. 2017. https://www.bjs.gov/index.cfm?iid=2237&ty=pbdetail. Accessed April 2017.
6. Carlone, H. B. and A. Johnson. 2007. Understanding the science experiences of successful women of color: Science identity as an analytic lens. *Journal of Research in Science Teaching*. 44(8): 1187–1218.
7. Chapman, T. K. and N. Hobbel. (Eds). 2010. *Social Justice Pedagogy Across the Curriculum: The Practice of Freedom*. New York, NY: Routledge.
8. Desmond, M. 2016. *Evicted: Poverty and Profit in the American City*. New York, NY: Crown Publishers.

9. Frankenstein, M. 1983. Critical mathematics education: An application of Paulo Freire's epistemology. *Journal of Education*. 165(4): 315–339.
10. Gutstein, E. and B. Peterson. 2013. *Rethinking Mathematics: Teaching Social Justice by the Numbers*. Madison, WI: Rethinking Schools.
11. Harrison, L. 2015. Teaching social justice through mathematics: A self-study of bridging theory to practice. *Middle Grades Review*. 1(1): 1–12.
12. Katz, B. 2017. Supremum/supremacy. https://blogs.ams.org/inclusionexclusion/2017/05/04/supremumsupremacy/. Accessed 1 November 2017.
13. Kelly, H. 2016. Why everyone is cracking down on Airbnbhttp://money.cnn.com/2016/06/22/technology/airbnb-regulations/index.html. Accessed 1 November 2017.
14. Killpack, T. L. and L. C. Melon. 2016. Toward inclusive STEM classrooms: What personal role do faculty play? *CBE Life Sciences Education*. 15(3): 1–9.
15. Lee, Y. H. 2015. Yes, U.S. locks people up at higher rate than any other country. https://www.washingtonpost.com/news/fact-checker/wp/2015/07/07/yes-u-s-locks-people-up-at-a-higher-rate-than-any-other-country. Accessed 31 December 2018.
16. MacMillan, D. 2016. Airbnb Sues San Francisco to block rental registration law. *Wall Street Journal*, June 27. https://www.wsj.com/articles/airbnb-sues-san-francisco-to-block-rental-registration-law-1467081805. Accessed 31 December 2018.
17. Muessig, B. 2015. The Airbnb effect. http://www.sfchronicle.com/airbnb-impact-san-francisco-2015/#1. Accessed 1 November 2017.
18. Pfaff, J. F. 2017. *Locked In: The True Causes of Mass Incarceration and How to Achieve Real Reform*. New York, NY: Basic Books.
19. San Francisco Rent Board. 2016. http://sfrb.org. Accessed 1 April 2017.
20. San Francisco's Anti-displacement Coalition. 2015. San Francisco's eviction crisis 2015. http://www.antievictionmappingproject.net/FINAL%20DRAFT%204-20.pdf. Accessed November 2017.
21. Sentencing Project. 2017. http://www.sentencingproject.org/. Accessed 1 April 2017.
22. University of Albany. Source book of Criminal Justice Statistics. http://www.albany.edu/sourcebook/. Accessed 1 April 2017.
23. US Census Bureau. 2017. American Community Survey. https://www.census.gov/topics/income-poverty/income-inequality.html. Accessed 26 April 2017.
24. Young, M., D. Lambert, C. Roberts, and M. Roberts. 2014. *Knowledge and the Future School: Curriculum and Social Justice*. New York, NY: Bloomsbury Publishing.

Math for the Benefit of Society: A New MATLAB-Based Gen-Ed Course

Paul Isihara, Edwin Townsend, Richard Ndkezi and Kevin Tully

Abstract: Responding to a call for national reform of mathematical education, as well as a college-wide revision of general education (GE) requirements, we describe a new entry-level, GE course focused on the humanitarian utility of mathematics. This includes a detailed overview of how we taught the course using a Humanitarian MATLAB Lab Manual developed collaboratively with undergraduate students within an applied math program. A *PRIMUS* edition of this manual with complete MATLAB materials for a GE course is included as an Appendix for interested readers. A variety of social justice issues including peace-building after modern civil wars, continued cancer risk after Chernobyl, gang reduction, various dimensions of human trafficking, the use of a tractor and herbicide in subsistence farming, equitable resource distribution during the Syrian refugee crisis, and access to HIV vaccines are included in this special edition of the manual.

1. INTRODUCTION

The status quo is unacceptable. Such is the consensus of leaders spear-heading reform in college-level mathematics education. In May 2015, an extraordinary collaboration of five major American mathematical

Color versions of one or more of the figures in the article can be found online at www.tandfonline.com/upri.

organizations resulted in a "common vision" for excellence [3], which includes at its core, application of advanced mathematical software, descriptive statistics, and mathematical modeling. Their report twice specifies MATLAB as a beneficial technical tool for students to acquire, as it is widely used in many scientific and engineering fields. In keeping with this vision, as well as our college-wide revision of general education (GE), a team of faculty and undergraduate students within an applied math program designed materials for a new entry-level course to show how mathematics applies to important humanitarian issues. Having now taught this course twice (Spring and Fall of 2017), we present a detailed course overview (Section 2), assessment of student learning (Section 3), and complete MATLAB course materials (available on *PRIMUS*'s on-line Appendix).

2. COURSE GOALS

In designing a new quantitative-reasoning, GE course on Math for the Benefit of Society, three main goals emerged:

- (Goal 1) heighten student awareness of a variety of humanitarian issues;
- (Goal 2) show how a variety of mathematical models provide insight into these issues; and
- (Goal 3) develop basic proficiency in MATLAB and writing lab reports using the free, on-line LaTeX editor Overleaf.

2.1. Heighten Student Awareness

To support Goal 1, we included in the course materials a number of quotes to highlight the humanitarian need. For example, we prefaced Classroom Demo 3 on Chernobyl cancer risk with this quote:

> Victor Khanayev, a surgeon in the Russian district of Novozybko near Chernobyl, reported recently that many poor people have no choice but to eat food that is contaminated, including food from their own gardens. Halina Chmulevych, a single mother of two living in a village in Ukraine's Rivne region, was cited in the report as saying she sometimes had little choice but to feed her children contaminated food. 'We have milk and bake bread ourselves that, yes, is with radiation,' she was quoted as saying. 'Everything here is with radiation. Of course it worries me, but what can I do?' [2]

In class, short video clips about trafficking and other humanitarian issues were discussed before beginning MATLAB lab-related work as a way to ground the mathematical models in reality. For example, we showed an excerpt from Secretary of State John Kerry's speech upon release of the State Department 2016 Trafficking in Persons Report; a press interview of Andrew Forrest, head of the Walk Free Foundation, which produces the Global Slavery Index; a "homecoming" chapel message by social worker and Cairns University alumna Heather Evans; and in our last session on trafficking, a moving TedxDayton talk by youthful trafficking victim Catalleya Storm. Although use of videos is a viable method to heighten student awareness, an instructor with a passion for social justice will likely have additional ways to enhance this aspect of the course.

2.2. Introduce a Variety of Mathematical Models

Our approach to GE-level quantitative-reasoning was to use a variety of mathematical models in elementary exploratory data analysis covered by the MATLAB labs:

- pie and bar charts to highlight small tables of trafficking demographic data (Lab 1);
- scatter plots with simple linear regression lines to compare three trafficking indices: the U.S. State Department's 3-Tier classification, the European Union's 3P index, and the Walk Free Foundation's Global Slavery Index (GSI) (Lab 2);
- a sorting algorithm to rank governments by their ability to meet five response objectives to fight human trafficking, followed by a simple linear regression of the data for comparing the most correlated and least correlated pairs of objectives (Lab 3);
- a multiple linear regression model to predict trafficking prevalence (i.e., proportion of total population enslaved) using five vulnerability variables as inputs (Lab 4);
- a discrete dynamical systems model to measure the effectiveness of anti-slavery policies affecting the child and adult vulnerable and trafficked populations (Lab 5);
- standardized and normalized data to compare humanitarian disaster relief organizations, and then creating a multiple linear regression model which predicts the number of people served by the organization based on their number of full-time and number of volunteer staff (Lab 6);
- linear programming to optimize the cultivation method (ox vs. tractor, manual weeding vs. herbicide) given constraints on collective subsistence farming, and then performing a sensitivity analysis showing how market price affects optimal land usage (Lab 7);

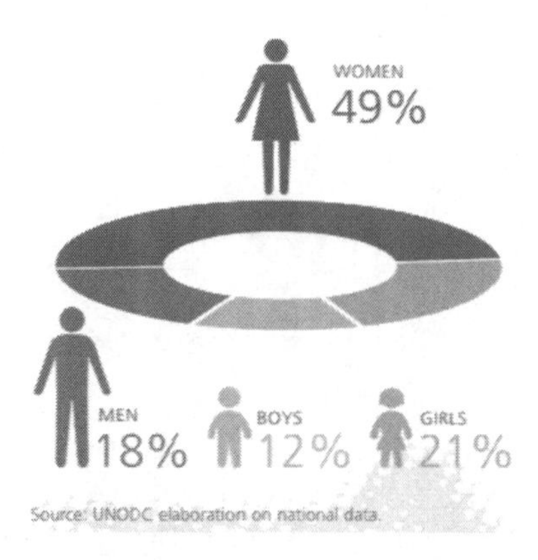

Figure 1. Age and gender of detected trafficking victims in 2011. The UNODC's 2014 Gobal Report on Trafficking in Persons is filled with visual displays of quantitative data, and gives permission to disseminate such graphs.

- Gini index methodology to determine equitable distribution of relief to Syrian refugees (Lab 8);
- multiple linear regression prediction of most violent typhoon positions and wind speeds, and a Monte Carlo simulation to assess the accuracy of these predictions (Lab 9); and
- an introduction to ODE models to determine steady state core gang and related populations (Lab 10).

These models provide various insights into the humanitarian problems being addressed. For example, very simple pie and bar charts in Lab 1 highlight the vulnerability of women and children to human trafficking, and majority of male perpetrators. In fact, the United Nations Office on Drugs and Crime's (UNODC)'s 2014 Global Report on Trafficking in Person [4] is filled with such graphical displays conveying information about trafficking in a clear and poignant way (Figure 1). Lab 1 has students chart the rise of forced labor as a second major form of trafficking, next to sex trafficking.

Early on in our project, we met with a cross-disciplinary group of faculty who were questioning the validity of the 2014 GSI [7]. Challenged to find an approach to the GSI validity question at a GE level, we developed Lab 2, which compares the GSI to two other important trafficking indices, the U.S. State Departments's 3-Tier ranking [5] and the European Union's 3P index [1]. During the introduction to the lab, we show how to create a scatter plot of (3P,GSI) values for each country identified by its 3-Tier ranking (Figure 2), observing a clear separation between the U.S. State Department's Tier 1 countries (most proactive in fighting trafficking) and Tier 3 countries (least proactive). Students then

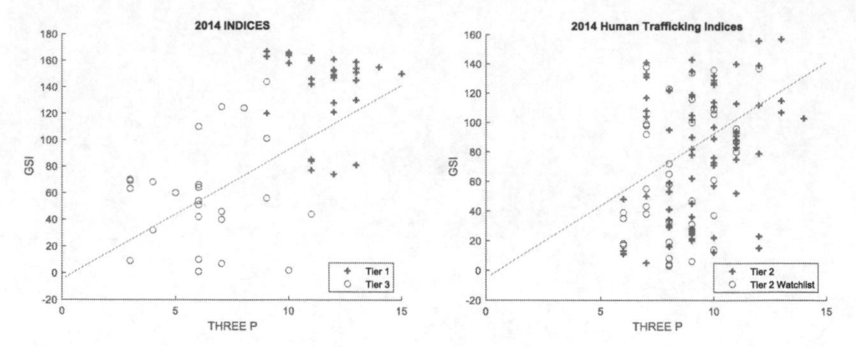

Figure 2. Left: A scatter plot of (3P, GSI) values with points designated by their Tier ranking shows a clear separation between Tier 1 and Tier 3 countries. Right: Separation is much less apparent for Tier 2 and Tier 2 watchlist countries. The regression line for the entire data set (all four Tiers) is included in both plots.

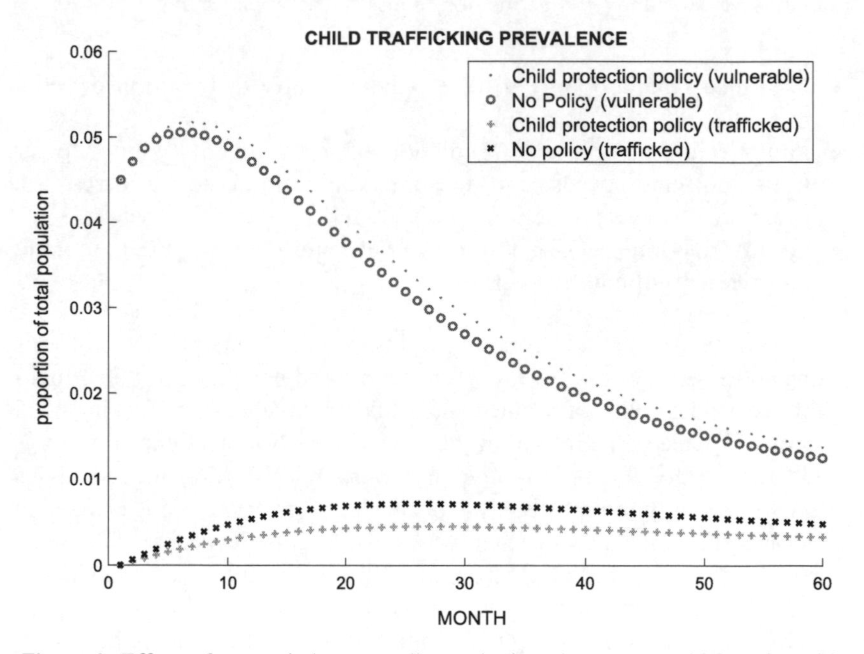

Figure 3. Effect of an anti-slavery policy reducing the rate at which vulnerable children are trafficked. Students were asked why the separation of the lower graphs is greater than the upper graphs. (A rate of change between two compartments may affect other compartments in a dynamical system.)

discover that the separation between Tier 2 and Tier 2 watch-list countries, which fall between the Tier 1 and 3 countries in their trafficking stance, is much less observable.

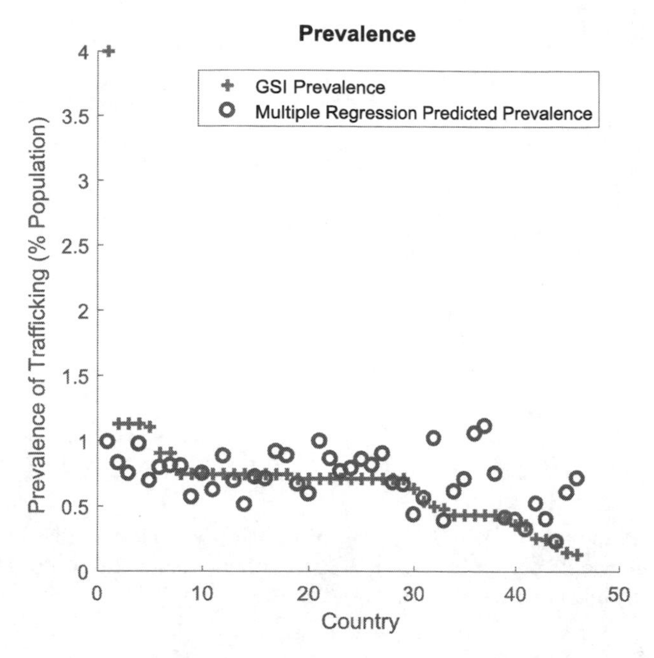

Figure 4. The clear outlier for sub-Saharan African countries is Mauritania (Country #1), with a GSI prevalence of 4%. Note the large residual error in the multiple regression prediction.

In engaging students to think about MATLAB output figures, quantitative reasoning skills may be demonstrated. For example, in Lab 5 we introduce a discrete dynamical system to keep track of safe, vulnerable and trafficked children and adults. For this simulation, we use relative proportions to measure the size of each subgroup, with the total population equal to one. The model shows the effect of anti-slavery policies, which would reduce or increase the rate of transition between two groups. For example, Figure 3 displays the effect of reducing the rate at which vulnerable children are trafficked. The difference in vertical spacing between the top pair and bottom pair of graphs indicates that an anti-child trafficking policy causes a bigger reduction in the child trafficked population than increase in the child vulnerable population. The model also indicates that the relative adult trafficked population will also be reduced, and the adult safe and adult vulnerable populations increased. Changing the transition between two compartments in a dynamical system may affect other compartments as well. (This idea is also explored in Demo 5 and the Appendix B Lab on Alzheimer's Disease therapy.)

Each lab concludes with one or more discussion questions, in most cases to reinforce the humanitarian issue. This section of the lab can

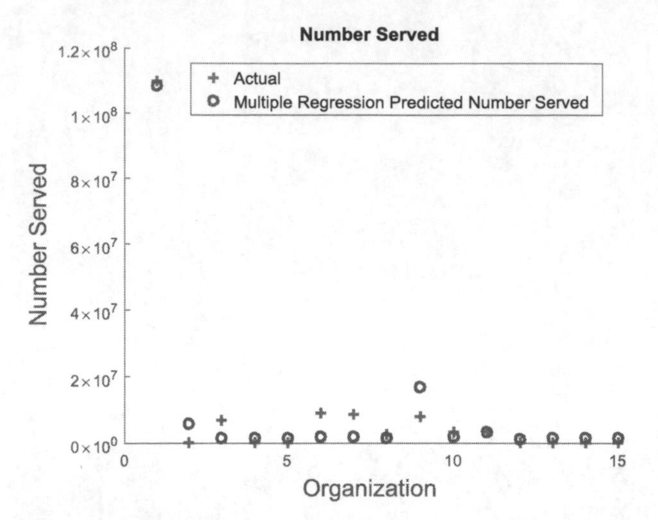

Figure 5. Although the IFRC (Organization #1) was a clear outlier in terms of number of people served by major disaster relief organizations, the regression model's residual error for IFRC was small.

easily be expanded, and so provides the instructor with a way to build into the lab additional teachable moments. This may include mathematical observations that connect ideas in previous labs. For example, a Lab 6 discussion question connects two data sets with similar structure, but there is a big difference in the multiple linear regression residual for each dataset's single outlier. The first dataset was introduced in Lab 4, in which country-level data for five vulnerability variables (see Lab4.xlsx, downloadable at https://github.com/pisihara/PRIMUS) are used to predict the GSI's reported trafficking prevalence [7] meaning, the percentage of a country's total population victimized by trafficking. For the sub-Saharan African region, Mauritania's GSI prevalence of 4% is far above all the other countries in the region. The residual error in the multiple linear regression for Mauritania is large, as indicated by country #1 in Figure 4. This error highlights the difficulty of acquiring accurate trafficking prevalence data and the GSI's reliance on regional extrapolation (countries deemed similar are assigned similar prevalences). About Mauritania, the GSI report [7] says:

> SOS Escalves, a reputable Mauritanian anti-slavery organization, as well as the BBC, and other reliable secondary sources, suggest that the proportion of the Mauritanian population enslaved is between 10% and 20%. However, no random sample survey information is available, no census has been performed in the country for some time (even the number of people in the total population is in doubt), and the government has been less forthcoming with demographic information, due in large part to the inter-ethnic tensions within Mauritania. Because

Table 1. Lab manual contents

HUMANITARIAN MATLAB LAB MANUAL *PRIMUS* EDITION

(*Continued*)

Table 1. (*Continued*)

of these caveats, the Walk Free Foundation retained the more conservative estimate used in the 2013 index.

A second data set with similar structure is shown in Figure 5. Even though the International Federation of Red Cross (IFRC) and Red Crescent Societies (IFRC) is the single clear outlier in the scatter plot of number served, the multiple regression residual error for IFRC is small. In contrast with trafficking data, the humanitarian organization data for the number of staff, volunteers, and numbers served (Lab6.xlsx downloadable at https://github.com/pisihara/PRIMUS) is easier to obtain and more reliable.

The material in the MATLAB manual displayed in Table 1 has been sequenced so that a mathematical method such as simple or multiple linear regression, introduced in an earlier lab, may be revisited in later labs. Instructors will need to introduce the underlying mathematical models during lectures, and may wish to add to the lab discussion questions to highlight or reinforce mathematical insights and connections arising from the MATLAB analyses.

2.3. Basic MATLAB Proficiency and LaTeX Lab Reports

To help GE students acquire MATLAB proficiency, we used a substantial portion of a Humanitarian MATLAB Lab Manual developed with the assistance of a large number of undergraduate math and applied math majors over a 2-year period. The Appendix contains a special edition of this manual for *PRIMUS* readers (see Table 1). This edition is tailored to the GE course, with our permission to freely use its contents for classroom instruction. In addition to a Beginner's Guide and nine pre-calculus level MATLAB labs, the *PRIMUS* edition includes several classroom demos and "mini-labs" to help students gain confidence in running MATLAB scripts and writing LaTeX reports before tackling the main labs.

The main MATLAB learning objectives are introduced in the seven classroom demos:

- DEMO 1: Knowing how to create and execute a MATLAB script, and where to find the values of the output data variables;
- DEMO 2: Reading in data from an Excel file and using a "for … end" loop to process the data;
- DEMO 3: Creating a figure using indexed MATLAB data variables, including axes labels and a figure legend;
- DEMOS 4-6: Using an auxiliary (user-defined) function and performing a sensitivity analysis on a key model parameter; and
- DEMO 7: Applying a MATLAB toolbox to analyze data.

These learning objectives are also found in the Beginner's Guide, six Minilabs, and 10 main MATLAB labs found in the *PRIMUS* Edition of the lab manual. This manual is designed to bring students to the level of being able to perform a sensitivity analysis using an auxiliary (user-defined) function. The auxiliary function takes key model parameters as inputs and uses an underlying mathematical model (e.g., multiple linear regression or linear programming) to compute a desired output. We found that GE students are able to employ a given auxiliary function as a "black-box." As a result, we have included in the *PRIMUS* Edition Demos 4-6 whose auxiliary functions are based on the three ODE labs found in the Manual's Appendix. We also added Lab 10 as a bridge to this Appendix.

In a GE class of 20–25 students, one would expect to find "non-math types" in the vast majority. To help facilitate student achievement, we suggest teaching this course in three stages.

- **Stage 1** (1–4 weeks): *Goal*: *Build a friendly and relaxed classroom atmosphere by considering topics of interest to the majority of the students and group activities at which they excel.* Our class met twice a week for 105 minutes each period. During the first weekly period we spent 30–45 minutes to introduce a discussion topic for the week. The remainder of the class was spent in groups of three or four students preparing a PowerPoint or poster presentation on various assigned aspects of the week's topic. For example, the topic "Humanitarian Organizations" could include group assignments to discuss the historical mission and current priorities of disaster relief and social justice organizations such as the IFRC/Red Crescent Societies and the International Justice Mission. The second weekly class period was devoted to the student presentations. In general, students gave excellent PowerPoint presentations. By giving specific assignments within a common theme, a broader overview and variety of each topic was achieved. Students were also required to critique each presentation, and written comments were collated and returned to each group. We kept the same groups for all presentations.

- **Stage 2** (2–4 weeks): *Goal*: *Ensure that no student gets left behind in transitioning to humanitarian* MATLAB *data analysis.* Before starting the actual MATLAB labs, we spent several weeks on MATLAB classroom demos, minilabs, and an introductory MATLAB Beginner's Guide. We repeatedly went through the same process as a class: (i) discussed and executed a short MATLAB script to get an output figure; (ii) saved the figure as a .png file and uploaded the script and figure to Overleaf; (iii) used a LaTeX template to create a brief report that included the output figure and MATLAB script. Most of the demos, Excel-related portions of the Beginner's Guide, and minilabs use humanitarian or social justice data. A few examples include peace-building after civil wars, cancer risk after Chernobyl, human trafficking, HIV/AIDS and Alzheimer's Disease treatment, pollution, and terrorism. Being shorter, demos and minilabs are easier to write than the main labs should additional topics be needed.
- **Stage 3** (7–8 weeks): Complete seven or eight MATLAB labs, without losing sight of the underlying humanitarian issues. Completing just one lab per week allowed ample time in our first weekly class period to: (i) discuss the humanitarian issue using relevant YouTube videos, Ted Talks, biographical anecdotes, etc.; (ii) explain step-by-step the mathematical model underlying the MATLAB computation (e.g., what is meant by the objective function and constraints in a linear programming problem?); and (iii) go through an example script line-by-line, so students could understand the main parts of the algorithm and the MATLAB commands needed to set-up and implement the mathematical model.

Every students was given a free PDF version of the MATLAB lab manual with the stipulation that they should purchase the student version of MATLAB available from Mathworks for about $50. Ideally, for an additional $50, students should also purchase the Simulink Student Suite of toolboxes. This Suite includes the optimization toolbox's function `linprog()` used for linear programming in Subsistence Farming Lab 7. (Students had difficulty acquiring MATLAB's 30-day free trial of this toolbox, so we allowed Lab 7 to be done in groups.) Demo 7, which requires a Toolbox not included in the Simulink Student Suite, may be used to illustrate how a new toolbox is acquired, and the fact that certain MATLAB functions will only run with their associated Toolbox.

Students brought their laptops to class, so our classroom became our MATLAB lab. After the introductory presentation, students worked individually on the lab at their desks, with a Teaching Assistant (TA) helping

Table 2. General course information

Item	Description
Credit	Our 16 week, four semester hour course met twice weekly for 110 minutes each session
Number Offerings	We taught one section in Spring 2017 and one section in Fall 2017
Enrollment	We capped enrollment at 25 and had nearly full enrollment both times the course was offered
Classroom Resources	We met in a regular classroom with a data projector and document camera, which were used to discuss MATLAB scripts. Most of the students purchased the student version of MATLAB and brought their laptops to class. As a result, the classroom was also our computer lab
TAs	There was a TA present at almost every class session, a tremendous help to debug scripts. The second time the course was offered, TAs were involved grading the labs, which were submitted as links to an Overleaf report
Prerequisites	This was an entry-level GE course with no formal prerequisites. In particular, no previous programming experience was assumed
Target Audience	The course met a quantitative reasoning GE requirement at a liberal arts college. The course was a mix of freshman through seniors, almost entirely non-STEM majors
Assessment	Students were required to submit a weekly MATLAB lab report written in LaTeX. There were no exams except for a final open-note exam in which the students were required to complete a brief lab report doing a simple MATLAB analysis of data they had not seen before

debug scripts and problems creating the LaTeX report. In all cases except for Labs 5 and 7, students wrote the scripts from scratch using the lab's example script as a guide. (Except for those used in the Beginner's Guide, most of the Example MATLAB Scripts in the manual may be downloaded by the instructor from https://github.com/pisihara/MATLABLABSFORPRIMUSEDITION). It was a tremendous help to have a TA assist with the debugging and grading of scripts. In most cases, students were able to complete the lab for the week by the end of the second class period. Students were allowed to submit their labs up to midnight the day of the second class period, and several needed this extra-time after class to complete their reports.

Our three stage approach was designed to give all students success in a gentle introduction to MATLAB and LaTeX. General information about the course is compiled in Table 2. The *PRIMUS* version of the MATLAB manual has sufficient material for such a GE course, and includes additional material for a more demanding and/or transition to a higher-level course.

3. STUDENT ASSESSMENT

Students were required to send links to their PDF Overleaf lab reports for a grade (example: https://www.overleaf.com/read/xkgqmwbkfqkb). A short, open-book final exam included an exercise to check that students had mastered the basic routine to: (i) modify a short script to make a MATLAB scatter plot of a dataset; (ii) upload the script and figure to Overleaf; and (iii) write a brief report using a LaTeX template. In keeping with our "gentle" introduction to humanitarian MATLAB modeling, our formal assessment was admittedly not very rigorous. We deemed it wise to begin with a course that most students can succeed, and gradually build more rigor in terms of both the mathematical content and MATLAB skills acquired. All concerned parties need patience to develop a truly effective course.

Much of the assessment of student learning was done informally, including gauging its humanitarian impact on students. For example, after watching Catalleya Storm's TedX talk, a student named Hannah said she had understood the need to be involved in fighting trafficking, but was not sure how to go about it. She said Catalleya's recommendation to not keep quiet, but to begin talking about the problem, was a good place for her to start. Similar encouragement to active involvement is possible when discussing the fact that Andrew Forest's daughter, after volunteering as a student in a Nepal orphanage, led her father to invest considerable resources to start the Walk Free Foundation, producers of the GSI.

Informal assessment also occurred when helping students debug their scripts. This gave a chance to check whether they understood the logic of the script and/or need for precise syntax. It was mutually gratifying when a script finally ran successfully. Similar to teaching quantitative reasoning about data, it is worthwhile to make an effort to help students understand the logic of a script or MATLAB command's syntax. For example, MATLAB has a similar syntax for simple linear (Labs 2 and 3) and multiple linear (Labs 4 and 6) regression. Increasing the number of predictors in the MATLAB syntax offers an intuitive context to transition from working with a function of one variable to multi-variable functions.

As mentioned earlier, students also showed understanding of the idea of a sensitivity analysis using an auxiliary function with one or more model parameters as inputs. Both "for … end" loops (first introduced in Classroom Demos 2 and 3 and an early part of the MATLAB Beginner's Guide), and auxiliary (user-defined) functions (first introduced in Demo

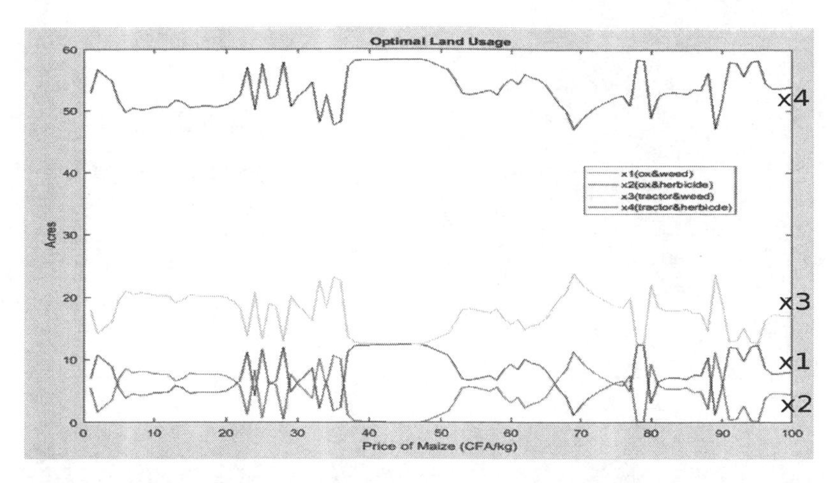

Figure 6. MATLAB 2016 gives turbulent solution behavior for the optimal number of acres using four cultivation methods for the market price range 25-35 CFA. (x1= acres ox-plowed and manually weeded, x2 = acres ox-plowed and herbicide sprayed, x3 = acres tractor-plowed and manually weeded, and x4 = acres tractor-plowed and herbicide sprayed.)

4 and the end of the Beginner's Guide), are an integral part of the MATLAB script writing. Having used these programming techniques repeatedly in earlier parts of the Lab Manual, students are then able to perform sensitivity analyses in Labs 5 and 7. This should not be done mechanically, but rather students should grasp that the index of the for loop determines the range of values of a key parameter, and a function is called in each iteration to determine a quantity of interest dependent on that parameter value. For example, in Lab 7, the auxiliary function outputs the optimal method of cultivating land (ox vs. tractor, manual weeding vs. herbicide) given the input market price of Maize.

When we first wrote this lab using MATLAB 2016, although the advantage of access to both a tractor and herbicide was clearly demonstrable, we were intrigued by the turbulent nature of the optimal solutions and switching between manually weeding and herbicide for ox-plowed acreage (Figure 6). One of the TAs for the GE course was puzzled that MATLAB 2017 gave a constant solution (Figure 7).

Dr. Pieter Mostermann, a senior research scientist at MathWorks (the company which produces MATLAB), responded to our question as follows:

> In R2016a, the default algorithm for linprog is "interior-point-legacy" whereas in R2017a the default algorithm is "dual-simplex". The latter default finds a basic solution (think of a corner of a polyhedron) to the

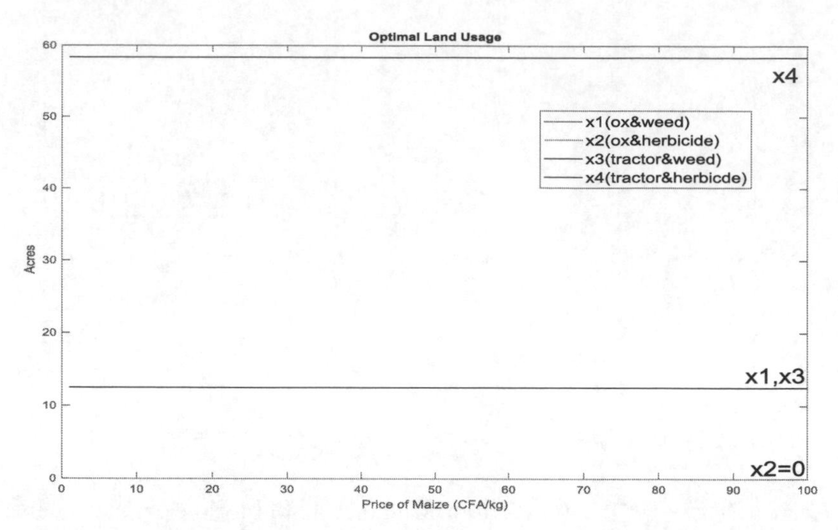

Figure 7. MATLAB 2017 indicates that the optimal land cultivation method is unaffected by the market price of maize. The MATLAB 2016 solution (Figure 6) is roughly the same in the market price range of 42-47 CFA/kg.

> linear problem whereas the interior point version could converge to any point on the face of the polyhedron that is parallel to the objective function gradient. It appears as though the problem posed by your students has multiple optimal solution points (i.e., face of the polyhedron) that give the same objective function value. So with interior point algorithm they will converge to one of the many solution points whereas with dual simplex they will consistently get one of the corner points. In a single optimal solution case, the R2017a version of the sensitivities makes sense. Note that the script solves many linear problems with the same constraints and a linearly changing objective function.

This reinforced our belief that interesting mathematics can arise at every level of instruction. Explained in simple terms, this example can help GE students acquire insight that numerical analysis involves much more than running a computer program.

Though the *PRIMUS* Edition of the lab manual contains considerably more material, we ended our GE course with this Subsistence Farming lab (#7). This lab is based on a linear programming model introduced by Dr. Maurice Vodounon, a Benin-born mathematician and United Nations Development Program (UNDP) consultant. Voudonon said:

> Very often there is little understanding of the problems that poor farmers struggle with in the village. Village organizations, farming cooperatives, and district planners do not have access to the organizations that decide which research projects receive funding. Yet there are many urgent problem areas in the everyday life of farmers that could benefit from such research. [6]

In developing our lab manual, we tried to contact Dr. Voudonon, only to learn he had passed away at a relatively young age. We have dedicated the *PRIMUS* Edition of the Humanitarian MATLAB lab manual in memory of Dr. Voudonon. He exemplified how mathematics can be used to address humanitarian concerns, a perspective we were excited to see many students express in a final one-page assignment to describe how their view of mathematics changed as a result of taking this course.

4. CONCLUSION

For those who may be considering offering this type of course, we hope our course description and complete set of MATLAB materials found in the on-line Appendix will help in your design and delivery of a meaningful course for both you and your students. Instructors who may wish to make the entire lab manual available to their students as well as obtain updates to the original *PRIMUS* Edition should check the URL https://goo.gl/KyB2dt. We also make available for students a number of data sets at the GitHub repository https://github.com/pisihara/PRIMUS and for instructors only, most of the Example scripts and auxiliary functions at another GitHub repository https://github.com/pisihara/MATLABLABSFORPRIMUSEDITION. As a GE math course, Mathematics for the Benefit of Society has a substantial technical component in terms of introducing students to a variety of mathematical models and MATLAB scripts in performing exploratory data analysis. Equally important is the goal to heighten our students' awareness of several important humanitarian and social justice issues. May the exhortation from Mother Theresa which concludes the Humanitarian MATLAB Manual serve as an encouragement to us all: *Let no one ever come to you without leaving better and happier.*

ACKNOWLEDGEMENTS

The authors would like to thank *PRIMUS*'s editorial review team, Stephen Kennedy, the Wheaton College (IL) Summer Research Program and Alumni Association, and the Council for Christian Colleges and Universities for their help and support of this project.

References

1. Cho, S. 2015. Report on the 3P Anti-traficking Policy Index 2014. http://www.economics-human-trafficking.net/mediapool/99/998280/data/Report_3P_Index_2014.pdf. Accessed 23 March 2016.

2. Reuters. 2016. 30 years after Chernobyl, food still radioactive, Greenpeace tests show. http://www.japantimes.co.jp/news/2016/03/09/world/science-health-world/30-years-chernobyl-food-still-radioactive-greenpeace-tests-show/#.WCzINC0rLcs. Accessed 18 November 2016.
3. Saxe, K., L. Braddy, J. Bailer, R. Farinelli, T. Holm, V. Mesa, U. Treisman, and P. Turner. 2015. *A Common Vision for Undergraduate Mathematical Sciences Programs in 2025*. Washington, DC: Mathematical Association of America.
4. United Nations Office on Drugs and Crime. 2014. Global report on trafficking in persons. https://www.unodc.org/documents/data-and-analysis/glotip/GLOTIP_2014_full_report.pdf. Accessed 23 March 2016.
5. U.S. Department of State. 2015. Trafficking in persons report 2014. http://www.state.gov/documents/organization/245365.pdf. Accessed 23 March 2016.
6. Vodounon, M. Mathematical models can make farmers more efficient. http://web.mit.edu/africantech/www/articles/MathFarming.htm. Accessed 2 December 2016.
7. WalkFree Foundation. 2015. The Global Slavery Index 2014: Methodology. https://www.walkfreefoundation.org/news/resource/the-global-slavery-index-2014/. Accessed 19 October 2018.

Using Graph Talks to Engage Undergraduates in Conversations Around Social Justice

Alison S. Marzocchi, Kelly Turner and Bridget K. Druken

Abstract: We present a mathematical activity called *graph talks* as a new pedagogical routine to intertwine social justice issues and mathematics. Adapted from *number talks*, graph talks involve students analyzing, interpreting, and discussing real-life data represented in graphs. Graphs may be strategically selected to both highlight a relevant social justice issue while also reinforcing the mathematics content of the course. We report on experiences using graph talks in undergraduate mathematics content courses for future teachers in the USA, and provide examples of the undergraduates doing mathematics while analyzing the social justice context of the graphs.

1. INTRODUCTION

In this article, we describe a new pedagogical routine called *graph talks* that is used to engage undergraduate students in connecting issues of social justice to mathematics content. A graph talk involves projecting a meaningful and relevant graph on the board as students enter the classroom and engaging in a 5-minute whole-class discussion of the mathematics alongside the social/environmental/economic implications of the graph. At first, it may seem as though conversations around social justice fit better in other disciplines, such as political science, criminal justice, social work, or humanities. However, we argue that undergraduate mathematics classrooms can be fruitful and appropriate venues for sparking such conversations. According to Leonard et al. [9] bringing issues of social justice into the mathematics classroom "can offer opportunities for students to learn mathematics in ways that are deeply meaningful and influential to the development of a positive mathematics identity." Unfortunately, school mathematics content is often taught separated from real-world experiences and authentic contexts, and is not connected to students' lives beyond the classroom [3]. However, opportunities exist to connect mathematics to issues of social justice, particularly through analyzing graphs.

It is crucial for our undergraduate students to participate in social justice conversations, and mathematics classrooms can be a productive space for engaging in these issues. The sample student comments below suggest that undergraduate students are willing and eager to connect mathematics to issues of social justice. When asked about whether issues of social justice have a place in the mathematics classroom, undergraduates who were future teachers enrolled in a mathematics course in which graph talks were frequently implemented, responded (note: all names are pseudonyms):

> **Stephanie:** Math helps us to quantify inequalities. Since we are future teachers, if you give us a number we will try to fix things. Where do we see inequalities and how do they manifest? What social programs can we put in place to help make it an equal playing ground for people?

> **Portia:** When we do graph talks, you see inequities, like fewer people of color graduating high school, going to college, and graduating from college. It makes you think about why that is ... Are they from different communities that don't have the resources to help people graduate high school and enroll in college? And you can show these graphs to people who maybe don't think there's an issue of race, you can show it with numbers and in picture form that there might be a discrepancy.

> **Amber:** By doing graph talks and learning how to read graphs, we can actually understand and come up with theories based on what we see. Most people read the numbers or look at the image and that's it - they don't think about it. Talking about social justice in math class can help us understand and develop theories.

In this paper, we first provide relevant background research on teaching mathematics for social justice and describe the theory behind graph talks. We then illuminate how graph talks could be used in undergraduate mathematics courses to link mathematics content and social justice topics. We conclude with examples from our own context to demonstrate how students communicate about mathematics and issues of social justice through graph talks. As suggested by the student responses above, we hope to suggest the power of graph talks to empower students with a sense of agency over issues that are important to them.

2. WHAT ARE GRAPH TALKS?

The pedagogical routine of *graph talks*, described in greater detail below, brings social justice awareness into mathematics classrooms by beginning class with a 5-minute conversation around a purposefully selected graph representing real-life, current, and meaningful data. The selected graphs often involve issues of social justice, which provide a foundation for deepening mathematical thinking in relevant contexts.

2.1. Background of Graph Talks

It is increasingly important to provide undergraduate students with opportunities to engage in conversations around social justice. Mathematics is a powerful, but perhaps underused, tool for students to question and analyze these issues. In a fairly recent educational shift, many teachers and researchers are focusing on cultivating a classroom environment centered around teaching mathematics for social justice (TMSJ) [1, 5, 6, 9, 10, 13]. Through TMSJ, "mathematics is used to teach and learn about issues of social injustice and to support arguments and actions aimed at promoting equitable change" [1]. With this lens, mathematics can be viewed as a tool to examine the lived worlds of the students, and to empower students to understand and question the conditions of their lives [6, 13]. Gutstein [6] sees TMSJ as having three components: "helping students develop socio-political consciousness, a sense of agency, and positive social and cultural identities."

Under ideal conditions, teachers at all levels would implement elements of TMSJ in their classrooms in a multitude of ways. However, research has shown that it is difficult, and often takes many years, for teachers to completely shift their pedagogy towards TMSJ [1, 5, 6, 9]. Some teachers have reported a discomfort in taking up mathematics problems with a "non-mathematical" context, as they are not "experts" in other fields [12]. However, Pfaff [12] asserts that:

> not being experts in areas other than mathematics should not dissuade us from having students consider non-mathematical aspects of problems, since it is in these non-mathematical aspects that students see the value of mathematics.

We suggest that the pedagogical routine of graph talks may be a feasible starting mechanism for bringing elements of TMSJ into the classroom, as teachers begin to build their comfort in linking mathematics to social justice.

GRAPH TALK
GUIDING QUESTIONS

- What is your first impression of the graph? What is the first thing you notice?
- What is the topic of the graph? What do the *x*- and *y*-axes each represent?
- What do you think is the purpose of this graph?
- Does this graph convey any trends? Are there any shifts?
- Did anything surprise you in this graph?
- If you had to make a prediction based on this graph, what would you predict?
- Do you think this graph is "fair" or is it misleading?
- What are the implications of this graph? What can/should we learn from it?

Figure 1. Poster of guiding questions that can be used in a graph talk.

Graph talks bear resemblance to, and were inspired by, a pedagogical technique known as *number talks*. Number talks are short pedagogical routines that allow students to mentally solve computations and discuss solutions with classmates [11]. Number talks have been used in K-12 classrooms and university settings. One example of a number talk is to mentally compute the solution to 5×18 [7]. Some solutions to this task include: (i) breaking apart 18 to 10 and 8, and finding five copies of 10 and add to five copies of 8; (ii) finding the sum of twenty fives instead of 18 fives, then removing two groups of five; or (iii) using factors and the associative property to think of $5 \times 2 \times 9$ as 10×9. By inviting students to mentally compute the solution to an operation, students can see that there are multiple ways to solve a task, and that only knowing the standard algorithm might limit their mathematical knowledge. One study found that frequent number talk routines had a positive effect on students' abilities to do mental computations and to use problem solving

The Higher Education Pipeline, By Race/Ethnicity
Percent of public high school graduates, 2011-2012 school year
59% White
18% Hispanic
16% Black
6% Asian
1% Other
Percent of 18-24 yr olds enrolled in college, 2012
58
19
14
7
2
Percent of 25-29 yr olds with a bachelor's degree or higher, 2012
69
9
9
11
2
Note: Hispanics are of any race. Whites include only non-Hispanics. For the high school graduate and bachelor's degree attainment figures, blacks and Asians include only non-Hispanics. For college enrollment figures, blacks and Asians include both Hispanics as well as non-Hispanics. "Other" includes small groups such as American Indians and those identifying as multiracial.
Source: U.S. Department of Education, National Center for Education Statistics; Pew Research Center tabulations of the March 2012 Current Population Survey Integrated Public Use Micro Samples (IPUMS); October 2012 Current Population Survey
PEW RESEARCH CENTER

Figure 2. Graph for a graph talk on degree completion by race/ethnicity.

strategies [8]. We see graph talks as an adapted version of number talks that provide a rich venue for linking mathematics content with issues of social justice.

2.2 Graph Talks in Action

To enact a graph talk, a real-life, current, and meaningful graph is projected as students enter the classroom (see Figures 2 to 6 for examples). A norm must be established such that students enter the room, take their seats, immediately look at the graph, and start conversing with those around them. In our mathematics classes, for future teachers of mathematics attending a large, public, southwestern institution known to be both

a Hispanic-serving and an Asian American and Native American Pacific Islander-serving institution, students are provided with a handout of "Graph Talk Guiding Questions" with an identical poster hanging in the room (see Figure 1). As indicated by the title of the handout, the questions are a *guide* and the students need not adhere strictly to the questions nor answer all of the questions. Instead, the guiding questions are meant to spark conversation. Instructors may benefit from modifying the guiding questions to better fit the content of their courses.

When class begins, the instructor initiates a student-led whole-class discussion around the graph. In our classes, we implement a discourse technique we call *chain chat* in which a student contributes to the discussion by building on a previous student's contribution. For instance a student might comment, "I can see why Stephanie thinks we might conclude ___ from the graph, but I actually feel we need more information because ___", or another may say, "I agree with Portia that ___ and I believe this may be because ___." During the discussion, students do most of the talking while the instructor serves to facilitate the conversation. This gives students important practice in engaging in conversations around social justice. We often find it helpful to scribe students' contributions on the board as they speak, as it serves to: (i) organize the students' contributions and record what has been said; and (ii) help instructors listen when tempted to steer the conversation. Graph talks are meant to provide opportunity for *students* to connect mathematics to issues in *their* world.

After approximately 5 minutes, the graph is put to rest and the daily lesson begins. We perceive that students' minds are now alert and primed for mathematical thinking. We believe this to be the case because students are engaged in thinking and conversing around mathematics content, prepping their minds for the rest of the mathematics lesson. At times, we are able to intentionally select graphs that directly align with the mathematics content discussed that current week. For instance, following a lesson on box and whisker plots, we intentionally selected a box and whisker graph for the subsequent graph talk. However, even if the selected graph is not directly aligned with the daily math content, we still sense that starting the class with a less formal mathematical conversation is a productive warm up before transitioning to the formalized mathematics content for the day. Additionally, frequent use of graph talks to start the class gives opportunity to convey an important over-arching theme week after week: that mathematics can be used as a tool for understanding issues of social justice.

Other researchers have found that some teachers report feeling overwhelmed by standardized test requirements and a dense mathematics curriculum. This may inhibit teachers from bringing issues of social justice into their classrooms for fear that there is not enough time [10]. Fortunately, graph talks occupy only 5 minutes of class time during a few lessons each month. From our experience, providing students with an engaging context around which to have a mathematical conversation as they enter the classroom helps students to productively begin class. Teachers may earn back those five minutes with students who are warmed up and ready to engage in the mathematics content. If a teacher cannot free up the time, they may consider implementing a *graph of the week* homework assignment, as described below by the second author.

2.3. Origins of Graph Talks: Graph of the Week

The idea for graph talks was sparked by a recurring assignment called *Graph of the Week* (GoW) implemented by the second author (Turner) in her AP Calculus AB and BC and Math 5 classes with high school students in a predominantly low-income school district [14]. Turner began GoW over 15 years ago as a means for student exploration of the relevance and importance of mathematics that existed beyond the textbook and outside the classroom. Due in part to the internet's speedy expansion where statistical data is spread through various social media platforms, students needed skills to analyze and interpret graphs, so that they may become informed and proactive citizens.

GoW could take on the following format: On Monday, the GoW is introduced and discussed briefly for 10 minutes. During the discussion students take notes, pose questions, make inferences, and analyze the intentions of the graph. Students are then given time to write independently followed by a partner discussion and second writing opportunity to inspire new insights. Turner has found that the more time students have to discuss the topic and to share their observations in the classroom, the more writing, and thus thinking, occurs. Students turn in the completed assignment on Friday.

This assignment helps engage students and make them feel at ease talking about topics they may not have thought would be discussed in a mathematics classroom: crime, obesity, abortion, divorce, sports, depression, drugs, alcohol, and unemployment, to name a few. Additional benefits include getting to know the students on a personal level and fostering a sense of community within the classroom. Readers should be cautioned that the maturity level of the students can impact their initial willingness to engage in conversations around these topics. Instructors may need to initially provide more guidance in a class of less mature students. Topics may be carefully selected to build students' comfort. For instance, students may feel more comfortable discussing sugar consumption in different regions of the United States than they would discussing unemployment. Students' skills in conversing around issues of social justice should be gradually guided and constructed, just as we aid students in building their mathematics knowledge over the course of a semester.

Inspired by the results reported by Turner, Marzocchi and Druken (the first and third authors of this paper) decided to test a modified version of the activity in their own classes, which they call *graph talks*. In what follows, we share our experiences implementing graph talks approximately two to four times a month in two undergraduate mathematics content courses for future teachers at a large public university of predominantly first-generation-college students.

3. EXAMPLES OF GRAPH TALKS

In this section, we share three purposefully selected and implemented examples: two *graph talks* and one GoW. For each, we provide the selected graph and rationale for its selection along with a reconstructed sample classroom discussion surrounding the graph. For each graph talk, one author facilitated the graph talk while the other author observed and took field notes. The roles were reversed and the graph talk was repeated in another section of the same course. Graph talk discussions were then reconstructed based on revisiting the field notes, viewing photographs of the scribed whiteboard, and debriefing conversations between instructors. Though some student comments are not direct quotes, discussions are illustrative recreations of the conversations surrounding the selected graphs.

Before we share our examples, we note the guiding question from Bartell's study of teachers learning to teach mathematics for social justice [1]. She asks: "In teaching mathematics for social justice, how does one negotiate the two goals of both *mathematics* and *social justice*?" A noted difficulty in bringing issues of social justice into the mathematics classroom is balancing content and context. In fact, researchers have found that teachers struggle to manage both and often default to discussing the two separately [1, 5, 10]. Garii and Rule [5], who examined student teachers' social justice mathematics lessons, found that "the integration of social justice and academic content was incomplete and, ultimately, the lessons themselves focused on either issues of social justice concern or academic content but not both."

We have noticed a similar dichotomy between targeting content and context in our own graph talks; students will discuss mathematics or social justice, and less frequently, the two concurrently. Students often discuss the mathematics first and the social justice context second. We conjecture that students may feel more comfortable discussing mathematics in a mathematics class whereas they are likely not accustomed to discussing social justice issues in this setting. More work is needed to examine why some graphs prompt students to focus more on mathematics whereas other graphs inspire students to focus more on the context. We believe this to be a fruitful area for future study and encourage others to investigate the nature of this dichotomy. We wonder:

- How might graphs be selected so that conversations include both mathematics content and social justice context?
- What contexts are more productive for instructors new to TMSJ to use in their particular undergraduate math course?
- What kinds of support do instructors need to do this work?

In the sections below, we provide an example of each scenario: a graph talk that primarily reinforces mathematics content, a graph of the week that primarily highlights a social justice issue, and a graph talk which simultaneously focuses on mathematics and social justice.

Example 1: A Graph Talk Reinforcing Mathematics Content

Like all activities implemented in our mathematics classes, graph talks serve a primary purpose of increasing students' mathematical skills and understanding. Though some may fear social justice conversations take time away from "doing mathematics," we conjecture that the meaningful context of the graphs will enhance the mathematics content. Past researchers have found that students' understanding of mathematical content is strengthened by a context that is rooted in their world [1, 12]. Of course, the benefits of graph talks are maximized when the selected graphs fit within, build upon, and/or review course content. For instance, if an instructor notices many of her students getting the stem-and-leaf problem wrong on the last exam, she may choose a graph with a stem-and-leaf plot for her graph talk. If an instructor feels students could use a refresher on percentages, they may find a graph that uses percentages.

To share an example of a graph talk with a discussion centering more on the mathematical content than the social justice context, we turn to a graph on "The Higher Education Pipeline, By Race/Ethnicity" (see Figure 2). This graph was selected because the first and third authors teach in a majority-minority university nationally recognized for bachelor's degree completion for students of color, particularly Latinx students. For this reason, we believed this graph would be an engaging context for our students and would allow them to connect the mathematics content meaningfully. Additionally, we thought it might spark pride in their own university that works to support historically marginalized students in the education pipeline.

Imagine the graph (see Figure 2) projected on the board as students enter the room. Once the students settle, the instructor initiates a whole-class chain chat about the graph. The sample vignettes below were constructed based on the observing author's field notes and serve as examples of discussions that may emerge from this graph.

The graph talk starts with a conversation around misleading graphs and study design. This is not surprising as we had recently discussed this content in our probability, geometry and statistics course for future teachers:

Instructor: Who would like to start our discussion on this graph?

Jan: I would change a few things about the graph such as keeping the percent symbol on all three bars and also labeling the races and ethnicities on all three. I also wonder about the sample. What was the size? Who was surveyed? Where were they from?

Instructor: It sounds like you're thinking about our recent lessons on misleading graphs and noting instances of this graph that may be misleading or potential sources of biases. Great. I wonder what else we can draw from this graph?

Vick: I think the purpose of the graph is to show that if you are a high school student, you are going to go to college at the same rate as graduating high school despite your race. But if the data is collected all during the same year, there's no correlation between the two sets. It's suggesting high school graduation leads to college entrance leads to degree completion but all of the data was (sic) taken the same year.

Instructor: Oh, can someone else explain more about what Vick is noticing with the study design?

Sarah: I was thinking those numbers should be somewhat the same across the three bars—percentages of people enrolled in college and the percent finishing with a bachelor's degree. But like Vick said, if data was (sic) collected during the same year, maybe that's why the percentages aren't the same.

Portia: [Vick] means data was (sic) collected during the same year, so the people enrolled aren't necessarily the same group. It's not following one population, it's different groups of people. In a better design, the ones that got a bachelor's degree should have been the same as the ones who enrolled in college. In this case, the percentages don't necessarily have to match because it's not the same group of people.

Instructor: It sounds like we think a longitudinal cohort study design model may have been more appropriate for the claims that are being made.

Stephanie: Yeah, I think it's trying to show potential dropout rates but it's not starting with the same sample group and following them through their whole trajectory. I think it's misleading that it's not the same sample group.

The instructor then tried to prompt a conversation around social justice by asking, "What about when you're looking at this data about who has a bachelor's degree and who doesn't? Does anything jump out at you?" For instance, the instructor anticipated the students noticing that although 18% of high school graduates in 2011–12 were Hispanic, only 9% of 25–29-year-old college graduates that same year were Hispanic. However, it is noted that the students addressed this discrepancy by primarily discussing the mathematics content of percentages:

Maria: I noticed the percentage for White students increased [from the high school graduates bar to the bar showing 25-29 year olds who hold bachelor's degrees or higher] but for many of the other races the percentage decreased.

Cyndi: I saw that, too, and I was thinking if more White people came to college, then the percentages would decrease for the other groups.

Maria: But I don't think it's that more White people suddenly came to college. I think it has more to do with dropout rates. That would change the percentages.

Instructor: Can someone comment on Cyndi and Maria's conversation? Are the percentages shifting due to more students attending college or due to people dropping out of college? Or maybe both?

Amber: Did the White bar increase and cause the others to get smaller, or did the others get smaller and so the White bar increased?

Maria: Well, what I'm saying is that if more Black and [Latinx] students drop out, and almost all of the White students stay in, then the percent of White students goes up, even if it's still the same number of White students. A higher percentage of remaining students are White compared to the other groups.

Lastly, the instructor shed light on the underlying social justice issue more overtly by asking, "Did the third bar graph change anyone's initial claim about the graph?" Students commented:

Savannah: It shows that enrolling in college doesn't mean you will definitely earn the degree. And it shows that less people of color end up getting their degree.

Instructor: And how can you see that?

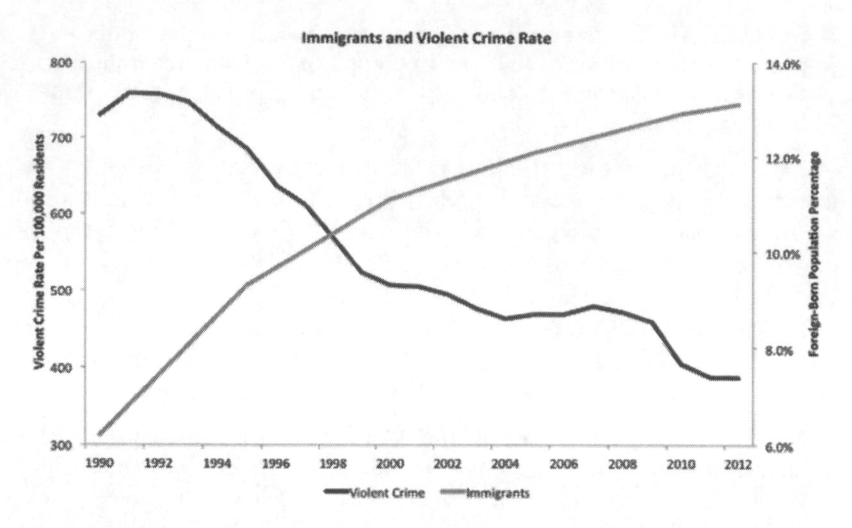

Figure 3. First graph from a GoW on immigration and crime.

> **Savannah:** By the lower percentages. From Hispanic it goes from 19% [in the second bar] to 9% [in the third bar].

> **Portia:** But if you look at the positive side of it, it's showing that maybe more people of color are enrolling in college than before. In the past, the percentage of Hispanics enrolling in college was probably less than the percentage graduating high school but now it's the same. It could be that the third level, even though lower percentages of people of color are finishing a degree, maybe over time we will see the numbers matching more closely. Enrollment improved, now graduation needs to improve.

It should be noted that students did make some mention of social justice implications of the graph. However, for this particular graph talk, contributions primarily centered around making sense of the mathematics of percentages.

Example 2: GoW Increasing Awareness of Social Justice Issues

Some may think that graph talks primarily fit in the context of a course on Statistics, but it is conjectured that graph talks could be used to enhance the mathematical content in a number of domains. Leonard et al. [9] provided examples of integrating social justice contexts in mathematics courses of various levels, including algebra, geometry, calculus, and fourth grade math. Not only will the context increase student understanding of the content, as discussed above, but the mathematical content may empower students to question the current conditions of their world.

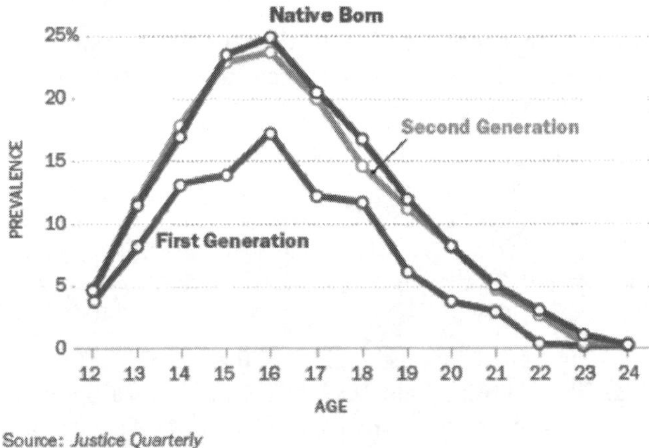

Figure 4. Second graph from a GoW on immigration and crime.

This next example shares a GoW which provided two graphs on "Immigrants and Crime" (see Figures 3 and 4). These graphs were selected due to their timeliness with national conversations around immigration. This GoW was enacted shortly after national accusations of immigrants as criminals and rapists. The second author seized this GoW as an opportunity to bring the national conversation into her classroom and to help her students, who are majority Latinx and largely immigrants or the children of immigrants, to critically investigate the claims made by national leaders.

Unlike the example above on degree completion, these graphs elicited student responses on the social justice issues more than the underlying mathematics. To start, the students wrote about the graphs for a few minutes. The teacher then asked the class, "Do you think socioeconomic status has something to do with crime rates?" A conversation similar to this ensued:

> **Aaliyah:** I think, sometimes, but not really. Most immigrants that come here don't start out rich. They are probably having low economic status, but it doesn't mean they commit crimes. They came here to make a better life for themselves, so they work low jobs or whatever they can get.

Ben: Yeah, the jobs they take aren't even all that great, but they will work the fields, or sell flowers on the street, whatever it takes to get themselves in a better place. They are honest workers, not criminals.

Clara: I know, and it's not like we're stealing jobs because I don't see White people working the fields.

The teacher then pressed, "Why do you think some people might believe that there *is* a correlation between immigrants and crime rates?" To which a student responded:

Damien: It's all the media's fault. They make it out to be racial all the time. It's like, if a crime was committed by a Black person, they will state the fact that he was Black. But, if a White person did it, they don't make reference to that. They let a guy like Brock Turner get out of prison sooner just 'cause he's White. But if a Black person did that, it would not be that easy.

Students continued this discussion at their tables. The teacher overheard things like, "Trump said that 'Mexicans are bringing crime, rapists… ' and that's so not true because it's not just Mexicans that do this, it's other Americans that were born here, and not just immigrants either," and also, "If people are well-to-do, they won't need to commit crimes like steal to get money."

This impassioned conversation then led students to further research the issue. They examined additional graphs and attempted to find evidence to back their claims. They discovered that immigrants are underrepresented in California prisons compared with the population overall and that men born in the U.S. are incarcerated at more than double the rate of immigrant men [4]. They also found that there was a small correlation between immigration and property crime, but only a slight one and that there are low levels of crimes committed in the lifetime of foreign-born individuals [2]. Though these graphs prompted more conversation around social justice issues than around the underlying mathematics, students' motivation to learn more about the issue was evidenced by their desire to do further research.

Example 3: A Graph Talk Simultaneously Discussing Mathematics and Social Justice

Though the examples provided above prompted students to focus more on either the mathematical content (see Figure 2) or the social justice issues (see Figures 3 and 4), this final example included conversations on both. We turn to a graph talk focusing on a set of three related graphs on "Wealth Distribution in the United States" (see Figures 5 and 6, the third graph did not have reprinting permission). Gutstein [6] implemented a similar context in a project with urban Latinx middle school

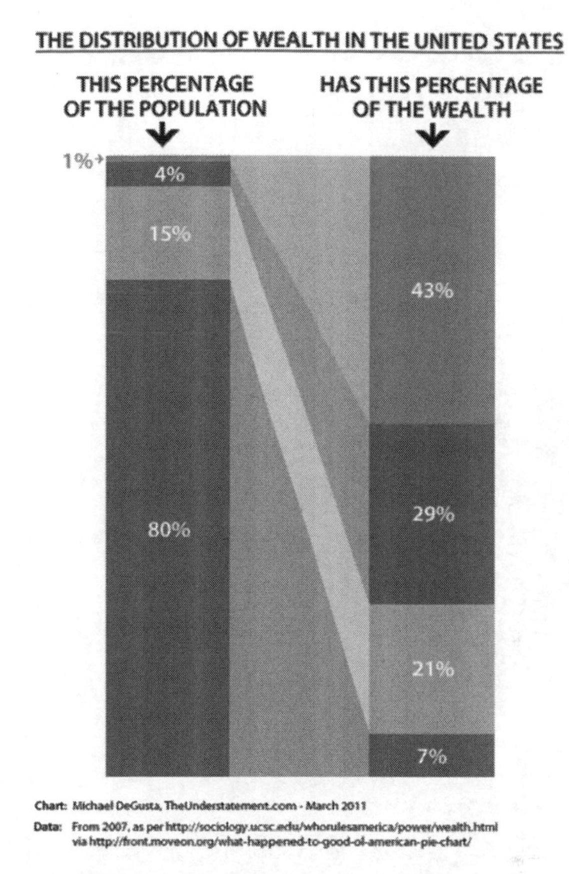

Figure 5. First graph from a graph talk on wealth distribution in the USA.

students. These graph were selected because the first and third authors believed that their students would benefit from seeing three different representations of the same data towards the end of a unit on statistics. They imagined this would spark conversation around the affordances and limitations of each representation, which would help them review for an upcoming midterm. Additionally, the authors were curious to see if and how the students would take up the social issue of wealth distribution.

For this graph talk, the three graphs were projected on the board as students entered the room, with identical printouts provided on each table for clarity, in case the three projected graphs were hard to read. One goal of the graph talk was for students to examine the amount of wealth held by different proportions of US citizens. For example, the graphs suggest that 1% of the US population holds 43%

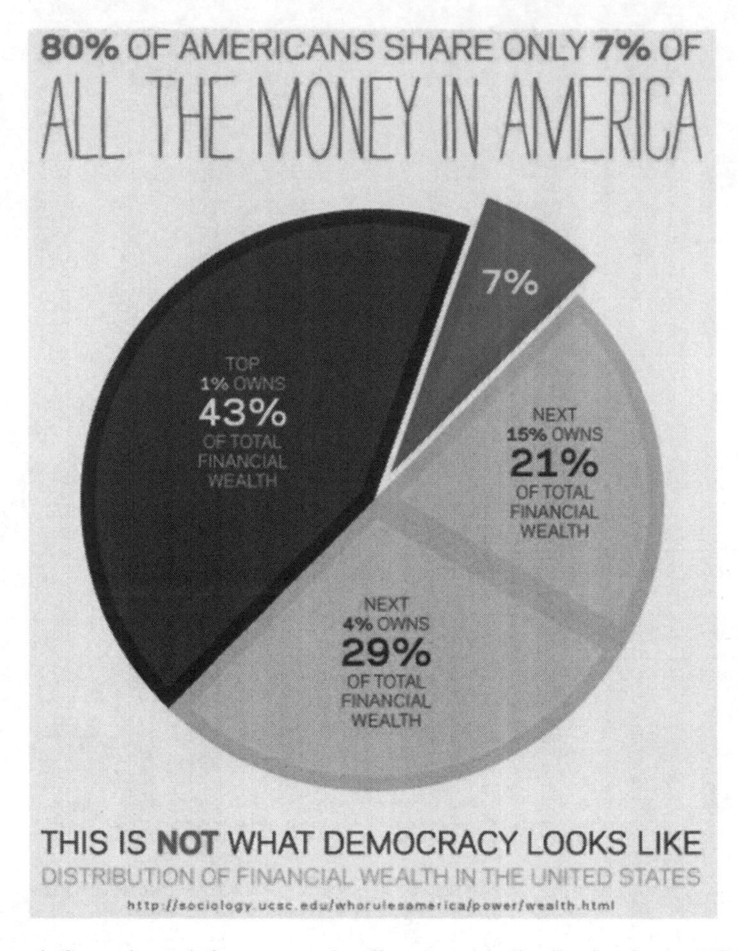

Figure 6. Second graph from a graph talk on wealth distribution in the USA.

of the country's wealth. Once the students settled at the beginning of class, the instructor initiated a whole-class chain chat about the graph. The sample vignettes below were constructed based on the observing author's field notes and serve as examples of discussions that may emerge from these graphs:

Instructor: Let's jump into the graph talk. What's your impression? What's the purpose of these graphs? Anything surprising you?

Deena: The graph with the little map on it, it says that little tiny red dot that you can't even see from here is worth 40% of the United States' wealth and yet 20% is that black area. The biggest sections of it are the smallest percents (sic), so it's kind of confusing. [note to readers: We did not have sufficient permission to reprint the map graph referenced here]

Instructor: It sounds like you are trying to understand what 20% and 40% are a percentage of. Can anyone clarify this?

Lucy: The percentages are talking about the population. So if you divide the US population up by wealth but represent wealth by land mass, it means that percent of the population would own that much space. So 40% of the population would use that much space worth of wealth. So they're using the United States map and chopping it up, instead of something like a bar or circle.

Kristen: The bar graph is the most impactful for me because it directly shows one percent of the US population owns that much wealth. It's more obvious that one percent has 43% of the wealth.

Instructor: What does this mean about 1% of the population having 43% of the wealth? Who can add on to this?

Fernando: The point of all three graphs is to show that a little bit of people hold a lot of wealth in the country. But it's not defined as what wealth is. Is it land? Cash? Stocks? Retirement? All? Because a lot of that accumulates over time and it is easier for rich people to get richer.

Aurora: I agree. The bar graph is a good representation of our capitalist society. It shows how in order for the top one percent to have that much wealth, the bottom 80 percent has less. There has to be this stratification.

The instructor then prompted approximately 20% of her students to stand up (six students) and stated that they represent the 20% of the population who, according to the graphs, possess 93% of the wealth. They sat down and the remaining 24 students stood up and were told that they represent the 80% of the population who, according to the graphs, possess only 7% of the wealth. The discussion continued:

Vickie: This is a problem! Because the 80% are probably the working class and the consumers. We need to spend money but how can we if we don't have money to spend? Rich people should be concerned about this, too.

Alicia: But the graph doesn't show dollar figures. What if the people in the bottom 80% still had enough money to be wealthy? I'm not saying they do, but we don't know it from the graph. Maybe it's not a problem.

Instructor: So the graph doesn't tell us how rich the rich are or how poor the poor are.

Alyssa: I feel like everybody has their own right to make money. Some make more. If they make more they have a right to buy whatever they want. If they aren't happy with the amount of money they have, then they should make more. Maybe it sounds mean, but that's why I'm back in school. I want more money, so here I am.

Instructor: I overheard someone mention the word "fairness" and I think this is an interesting word. Did anyone think about "fairness" and how this might come up in these graphs?

Amber: A lot of the times when talking about wealth distribution, it's related to tax policy. It's always the top 1% that pays the least amount of taxes compared to the lower class.

Anna: We talked about in the 80%, if the three of us were in 80%, we all make different amounts. So one person could be making a ton and then the rest are making even less than it appears.

Marcela: I want to know how this happened. Was it our heritage? Corruption?

River: Well, people with money can influence politics.

In this graph talk, it should be noted that students made mention of underlying mathematics content, such as percentages and construction of graphs, while also discussing the social justice issue of wealth distribution. This is a noted exception as students typically discuss either mathematics or social justice, but not both ideas simultaneously.

4. CONCLUSION

In this paper, we introduced *graph talks,* a new pedagogical tool for bringing conversations around social justice into the mathematics classroom. We believe that graph talks can provide opportunities to deepen mathematics concepts while simultaneously increasing students' knowledge of important world issues. It is important to note that the undergraduates included in this paper are future teachers of mathematics. This may be particularly powerful because increasing their comfort in, and ability to, connect mathematics to social issues could increase the likelihood that they will facilitate these conversations with their future students. This could help affect societal change in the ways that Bartell [1] and Gutstein [6] have hoped.

However, we believe that teaching mathematics for social justice has a place in other mathematics courses and that graph talks can be adapted to courses in addition to those teaching statistics to future teachers. For instance, rather than using potentially inauthentic or uninteresting textbook examples of graphs or data, an instructor can find data related to an engaging and important current issue. In fact, instructors might consider swapping out *any* graphs or data in the curriculum with those that pertain to social justice. In this way, TMSJ is not limited to statistics courses, but could be implemented in math for liberal arts, college algebra, pre-calculus, the calculus sequence, etc. It also suggests that collaborations with others involved in teaching social justice may be productive for mathematics instructors, applied mathematicians, sociologists, and economists, to name a few.

Openly conversing around issues of social justice can be considered a learned skill that may improve with reflective practice. As instructors who have implemented graph talks or GoW, we have experienced an increased sense of comfort in leading these discussions as we continue to implement them regularly. We conjecture that our students feel a similar sense of increased comfort. Although some of us were concerned that we would not be able to answer all the questions posed by students during a graph talk due to not being "experts," we recalled Pfaff's call for supporting experiences that engage students in topics that are of interest to them. When instructors get feedback from students that support continuing an activity and feel supported by their colleagues in doing so, it serves as reason for continuing. Future research could investigate instructors' and/or students' comfort with, and ability to, discuss social justice mathematics throughout the course of a semester. Moreover, we conjecture that connecting mathematics content to issues of interest to our students could increase engagement and learning opportunity. In fact, we encourage instructors to poll their students at the beginning of the semester to see which social issues are of interest. This data could help to guide the instructor in choosing relevant and meaningful contexts for unpacking mathematical content.

Through graph talks, students are empowered to use their mathematical knowledge to understand important issues. This aligns with Bartell's vision of education [1]:

> The purpose of education is not to integrate those who are marginalized into the existing society but rather to change society so that all are included. Thus, education should help students analyze oppression and critique inequities, highlight how these issues connect to their lives, and engage them in challenging those inequitable structures.

According to Gutstein [6], "helping young people develop a sense of personal and social agency can be an important step toward achieving equity." We suspect that graph talks increase students' awareness of socio-political topics, such as graduation rates and wealth distribution, and provides evidence from which students can take action in their own communities. Graph talks may serve as an implementable means for introducing social justice issues into the mathematics classroom and seeing inequities, quantifying inequities, and coming up with theories based on what we see, as suggested by Stephanie, Portia, and Amber in the Introduction.

REFERENCES

1. Bartell, T. G. 2013. Learning to teach mathematics for social justice: Negotiating social justice and mathematical goals. *Journal of Research in Mathematics Education*. 44(1): 129–163.
2. Bersani, B. E. 2014. An examination of first and second generation immigrant offending trajectories. *Justice Quarterly*. 31(2): 315–343.
3. Brown, J. S., A. Collins, and P. Duguid. 1989. Situated cognition and the culture of learning. *Education Research*. 18: 32–41.
4. Butcher, K. F. and A. M. Piehl. 2008. Crime, corrections, and California: *What does immigration have to do with it? California Counts*. 9(3):1–23.
5. Garii, B. and A. C. Rule. 2009. Integrating social justice with mathematics and science: An analysis of student teacher lessons. *Teaching and Teacher Education*. 25(3): 490–499.
6. Gutstein, E. 2003. Teaching and learning mathematics for social justice in an urban, Latino school. *Journal of Research in Mathematics Education*. 34(1): 37–73.
7. Humphreys, C. and R. Parker. 2015. *Making Number Talks Matter: Developing Mathematical Practices and Deepening Understanding, Grades 4-10*, Stenhouse Publishers.
8. Johnson, A. and A. Partio. 2014. The impact of regular number talks on mental math computation abilities. http://sophia.stkate.edu/maed/93. Accessed 15 February 2017.
9. Leonard, J., W. Brooks, J. Barnes-Johnson, and R. Q. Berry III. 2010. The nuances and complexities of teaching mathematics for cultural relevance and social justice. *Journal of Teacher Education*. 61(3): 261–270.
10. Nolan, K. 2009. Mathematics in and through social justice: Another misunderstood marriage? *Journal of Mathematics Teacher Education*. 12(3): 205–216.
11. Parrish, S. D. 2011. Number talks build numerical reasoning. *Teaching Children Mathematics*. 18(3): 198–206.
12. Pfaff, T. J. 2011. Educating about sustainability while enhancing calculus. *PRIMUS*. 21(4): 338–350.

13. Stinson, D. W., C. R. Bidwell, and G. C. Powell. 2012. Critical pedagogy and teaching mathematics for social justice. *International Journal of Critical Pedagogy*. 4(1): 76–94.
14. Turner, K. 2017. Turner's graph of the week. https://www.turnersgraphof-theweek.com/. Accessed 15 February 2017.

Critical Conversations on Social Justice in Undergraduate Mathematics

Nathan N. Alexander, Zeynep Teymuroglu and Carl R. Yerger

Abstract: This article explores how *critical conversations* engage undergraduate mathematics faculty in a community of practice that enhances their knowledge about teaching and learning mathematics for social justice. More broadly, critical conversations are defined as a cooperative learning strategy that can be used to identify, explore, and respond to various interests and issues situated across differing values and beliefs. We present a case study of a critical conversation that took place at a 2016 *Mathematics for Social Justice* workshop organized by a group of junior faculty. Participant reflections situate perspectives that can help novice and experienced instructors design conversations about teaching mathematics for social justice. Specifically, individual and group reflections highlight the importance of: (i) framing and reflecting on the conversation; (ii) exploring implications and content connections; and (iii) identifying barriers. Implications for faculty members and mathematics departments are provided.

1. INTRODUCTION

This article focuses on holding conversations that support faculty who are interested in developing conversations about incorporating social justice issues into undergraduate mathematics curricula at their institutions. We frame critical conversations as a practical component of professional development and utilize perspectives on teaching mathematics using a *social justice* lens. We open our discussion by situating critical conversations in a community of practice (CoP) [20, 30] and draw on critical theory [22, 29] to explore the various aspects of pedagogical and curricular development in teaching post-secondary mathematics for social justice. We then examine the components of a sample conversation focused on teaching mathematics for social justice. This conversation and participant reflections provide a case study that is used to frame some of the processes that can be used to organize similar critical conversations.

In this article, *critical conversations* are defined as a cooperative learning strategy that can be used to develop a participant's knowledge and examine the implications of different attitudes and beliefs about a shared topic on one's practices. Critical conversations on social justice in mathematics relates to the various ways that attitudes and beliefs about social justice might be explored and subsequently integrated into mathematics content courses. We focus our discussions on post-secondary mathematics, given the dearth of literature on teaching and learning mathematics for social justice at the undergraduate level.

Better understanding on how to generate critical conversations will allow practitioners and researchers to explore how structured conversations about teaching practices come as a result of engaging with different perspectives and attitudes. This article sheds light on *how* to foster critical conversations, both as a regular form of professional development and as a tool for the practical exchange of knowledge. This professional development is deemed *critical* because it increases one's opportunities to expand their knowledge by challenging privileged or mainstream systems of thought. Thus, critical conversations situate participants in a community of practitioners that will help participants frame and understand and also reflect and respond to critical issues of collective interest.

2. FRAMING CRITICAL CONVERSATIONS

Critical conversations represent a community strategy used to engage multiple perspectives. These conversations can help to improve knowledge about a topic or area of study. Jean Lave and Etienne Wenger define a CoP [20, 30] as a group of individuals who share a craft or profession (such as teaching). Within these communities of practice, knowledge can evolve naturally or it can be shared deliberately with the goal of situating perspectives and practices related to a focused area of study. Critical conversations represent a deliberate attempt to structure the exchange of knowledge and help frame and situate the learning exchange.

Wenger [30] expands on the idea of situated learning by theorizing that learning is an interactive social activity and not solely in a learner's head. Based on Lave and Wenger's [20] theory, the sociocultural processes that result from conversations (in the form of activities and other interactions) in a community of practitioners results in shifts from peripheral participation to more full participation and engagement. During these community interactions, new knowledge is gained and various professional practices can be shared and developed.

We propose critical conversations as one form of professional development based on the CoP framework. According to Lave in [20], CoPs help to foster engagement while enhancing participant's knowledge. These interactions begin with a common goal or purpose where the community builds trust to begin to exchange their knowledge. As a result of these exchanges, participants are able to explore their beliefs and understand how situated learning can be used to build on implicit knowledge through conversations and provide avenues to explore a wide range of beliefs [1].

Critical conversations also serve as an ideal setting to engage professors and teaching faculty in developing new practices or extending practices to their fields based on knowledge from other areas of study. The extensive research on teaching and learning for social justice in other domains, such as in K-12 education, represents one example of this knowledge exchange. Throughout this article, we explore how learning to teach mathematics for social justice is a community practice that begins with critical conversations. In the next section, we discuss some of the ways that critical conversations on social justice might develop.

2.1. Critical Conversations on Social Justice

Given that multiple definitions of "social justice" exist in education, critical conversations are important because they allow participants to explore how the terms *social* and *justice* might be used and defined. Connie North [24], building on Gewirtz's analyses of social justice in education [12], discusses how terms such as social justice are engaged across a wide variety of narratives and discourses that ultimately causes confusion based on the different perspectives held. North discusses the importance of "examining the tensions that emerge when various conceptualizations of social justice collide" as a way to "promote continued dialogue and reflexivity on the purposes and possibilities of education for social justice" [20, p. 507]. Critical conversations allow participants to identify these tensions and examine their implications for practice.

Holding critical conversations also has the potential to expose similarities and differences in logic and various assumptions held across participants' perspectives. Identifying these differences can surface opposing views of the local community and also of global social contexts. Thus, critical conversations not only help to localize what is meant by the term social justice, but also they can improve knowledge and engagement with issues of injustice overall. Critical conversations on social justice also equip participants with an understanding of their own perspectives on social issues in contrast with others' beliefs, and they allow faculty to exchange ideas and identify new goals for their own teaching and pedagogical practices.

As a result of identifying differences in perspectives and logic, we urge readers to keep in mind the importance of ensuring that multiple voices, as well as a diverse representation of participants, are invited into the discussions. Diversity in a CoP helps to ensure that a conversation is substantiated by the exchange of different ideas versus an agreement across ideas. To ensure this outcome, understanding the components of a critical conversation and the implications in a particular context is necessary. To help frame the components of a critical conversation, we will explore the case of a conversation focused on teaching mathematics for social justice.

We present a sample conversation to help readers make sense of a discussion and its tensions. We present reflections to help the reader understand the different positions of the participants. Since the implications of this conversation and the reflections depend on the local contexts of the reader, we end this article by providing an open-ended framework for readers interested in fostering critical conversations. In the next section, we explore the importance of diversity by engaging scholarly voices from across different fields. These diverse voices allow us to explore some foundations of critical theories and relate them to teaching mathematics for social justice.

2.2. Critical Pedagogy in Mathematics

As noted earlier, there are strong foundations in the literature on teaching and learning mathematics for social justice. These roots point to contemporary scholarship in the field of critical studies and specifically to the 1970s and 1980s in the work of Henry Giroux [13], Paulo Freire [8, 10], and Derrick Bell [3, 5], to name a few notable theorists. In these texts, *critical* generally refers to a practice of reflective assessments and criticisms of society and culture to understand the world. These theoretical foundations have impacted research scholarship across many areas of study, including the field of mathematics education. However, less work has been done with respect to undergraduate mathematics education.

Paulo Freire, the Brazilian educator and philosopher best known for his influential work, *Pedagogy of the Oppressed* [9], is widely considered to be the founder of the critical pedagogical movement. He defines *critical pedagogy* as a process in which we "become agents of curiosity, ... investigators, ... subjects in an ongoing process of quest for the revelation of the 'why' of things and facts" [11, p. 105]. From this lens, critical pedagogy and teaching for social justice centers practices that actualize the ways that education can be used not only for knowledge acquisition, but also as a tool to critique society and for reflective assessment. When education is used for reflective purposes, new perspectives and knowledge surfaces that may not have been present before. And as a method of critique, critical education allows students and teachers to experience the impact of knowledge and its relation to social justice.

In mathematics, critical pedagogy, and specifically teaching for social justice, can be used to teach quantitative literacy while also helping learners understand the role of mathematics in society. Social justice education exposes the ongoing struggles for equity and the liberation of historically oppressed persons. In teaching mathematics for social justice, educational practices are centered on the quantifiable and mathematical aspects of this knowledge. The interpretations of various critical theories and applications of these theories range in both concept and content; however, these roots provided substantive content to help others develop applications of critical pedagogy in mathematics [23, 27, 28].

Critical applications in mathematics, which produced a new body of work now known to the field as critical mathematics education (see Frankenstein [7] and Skovsmose [27]), situated social justice work in contemporary mathematical practices. This original body of work encouraged "teachers and students to develop an understanding of the interconnecting relationship among ideology, power, and culture, rejecting any claim to universal foundations for truth and culture, as well as any claim to objectivity" [28, p. 77]. Importantly, however, critical mathematics pedagogy should not be understood as a singular concept or static practice. Instead, it should be viewed as a set of practices in which collaborators use mathematics to critically engage the local contexts which surround them and in the world.

Moreover, different theoretical perspectives on teaching mathematics for social justice exist [17]. In Stinson et al. [28], for example, the authors explore the processes of teacher development with regard to applications of critical theory in mathematics. Much of the discussions in this article provide very useful examples of critical reflections on the various theoretical applications of social justice in mathematics, which they and others frame as TMfSJ (Teaching Mathematics for Social Justice [2]), and their implications for pre-service teachers.

However, whereas Stinson et al. [28] primarily focus on the contributions of critical theorists such as Freire, other mathematics education researchers examine intersecting critical perspectives from perspectives on critical theory. One recent example is the article by Larnell, et al. [19] that examines the capacity of theoretical perspectives on TMfSJ to address issues of race, and specifically other intersectional critical theories such as those developed by Derrick Bell [3] and others. Specifically, the authors examine the implications of critical race theoretical perspectives on learning to teach mathematics for social justice. Some of the important ideas raised in these discussions help to structure the needs in a community of practitioners focused on the various applications of social justice.

With a focus on practical classroom applications in K-12 mathematics education scholarship, Eric Gutstein [14, 15] has positioned teaching mathematics for social justice as a dialectical process of critical engagement. In this process, students and teachers collaborate to "read the world (understanding complex issues involving justice and equity) using mathematics, to develop mathematical power, and to change their orientation toward mathematics" [14, p. 37]. One example of practical curricular aides that convey the various perspectives on critical pedagogies in elementary and secondary mathematics has been outlined in key texts such as *Rethinking Mathematics*. This resource was edited by Eric Gutstein and Bob Peterson [16] and it contains a collection of 32 chapters that engage multiple perspectives that range across four key areas of concentration:

1. Historical perspectives and conversations on teaching and learning mathematics for social justice.
2. Activities, lesson plans, and lecture notes on teaching mathematics through the lens of social justice.
3. Discussions about the role of mathematics in social activism.
4. Resources, websites, datasets, and books on teaching mathematics for social justice for novice instructors.

The lesson plans and activities included in [16] build on the diversity of critical theories and present frameworks to help practitioners engage in the process of learning to teach mathematics for social justice. Examples of the lessons focus on discussing and calculating unemployment rates, examining neighborhood displacement, and using mathematics to design wheelchair ramps for improved and equitable access. This wealth of resources clearly exemplifies a solid foundation of applications that conveys a critical pedagogy in K-12 mathematics, but much less work has been done with regard to undergraduate mathematics teaching and learning.

Much of the current examples on teaching mathematics for social justice target K-12 students and teachers, and pre-service teachers and teacher educators. However, the social justice applications in these lessons tend to be more focused on elementary and secondary mathematics content. Importantly, critical conversations will play an important role among collegiate education scholars in helping to extend the lessons learned from K-12 practitioners to undergraduate mathematics education. Critical conversations will allow participants to take fundamental ideas in critical theory and the research in K-12 and apply them to the undergraduate mathematics context.

Critical conversations on social justice in undergraduate mathematics can provide teaching faculty with collaborative opportunities to generate relevant ideas, understand concepts, and develop activities from the critical mathematics education literature and apply them to the more advanced courses in undergraduate mathematics. We argue that engaging in these critical conversations is essential to developing one's knowledge prior to the work of designing and developing culturally and socially responsive curricula in higher education. This is especially important given that college faculty tend to work in silos as they develop courses and ideas for pedagogical practices. Furthermore, critical conversations in a CoP can help faculty deal with the nuances of generating applications of social justice in more advanced mathematics.

3. CRITICAL CONVERSATIONS IN UNDERGRADUATE MATHEMATICS

Our conversations about teaching and learning mathematics for social justice began with the development of a faculty workshop designed to explore applications of social justice in the undergraduate context. This 2-day workshop started with plenary lectures, all concentrating on different aspects of integrating mathematics and social justice, and continued in content-specific groups. In these small groups, considered as a community of practitioners, participants discussed how social justice content might be engaged in undergraduate settings and integrated into the mathematics curriculum. These discussions centered around elementary applications of social justice in a specific course or content area (e.g., calculus, statistics).

During their discussions, participants focused on creating activities, lesson plans, and projects. The goals of the lessons were to address complex issues of injustice—such as inequality in health, education, and wealth distributions, and the broad issues related to race and ethnicity and class and gender—and highlight applications to mathematics. The activities created in these small groups were intended to promote applications of mathematics with a focus on critical thinking and critical issues. As a result of these small groups, all participants engaged in different forms of critical conversations in a CoP.

The activities developed in the content-specific CoP groups allowed faculty to discuss ways that mathematics could be used to increase students' mathematical understanding while expanding their social and political consciousness, and also integrating students' diverse backgrounds and identities into the mathematics curriculum. In order to accomplish these tasks, faculty had to engage in critical conversations related to a specific area of the undergraduate mathematics curriculum. In the advanced statistics group, instead of choosing a single idea to highlight one specific social justice issue, participants decided that a discussion about the extent of different perspectives on the inclusion of social justice in statistics would be a valuable exercise. This particular CoP explored topics throughout an upper-level statistics course and sparked the critical conversation overviewed in this article.

This critical conversation (presented in the next section) focused on the extent to which one might integrate social issues in an advanced statistics course, and questioned how the time spent on issues of injustice could potentially shift the stated nature and goals of a course. Some participants felt that a significant amount of time should be spent on providing context to issues of social injustice. Others felt that this should not be the focus of a mathematics course, especially in more advanced courses that were not specifically placed in the curriculum to handle social justice issues.

As a result of these differences, key issues were explored to help frame applications of social justice in statistics. The disagreement noted in the previous paragraph can be linked as a part of a longstanding debate on the question of *mathematics content* versus *learner context* [26]. This theoretical issue is largely based on the practical differences found in the foundations of research on teaching mathematics. On one end of these foundations is the focus of mathematics content acquisition and, on the other end, is a focus on the context in which the mathematics is learned. This conversation and the associated small-group activity aided faculty members in thinking more concretely about different approaches to teaching undergraduate mathematics while also making sense of practices related to teaching mathematics for social justice.

3.1. A Sample Critical Conversation

To make better sense of how the workshop fostered critical conversations, in this section our discussions focus heavily on the interactions

Table 1. A misleading table

Race of Defendant	Death Penalty	Lesser Sentence
Caucasian	19	141
African American	17	149
Total	36	290

between participants of the 2016 Associated Colleges of the South (ACS) and Fitchburg State University (FSU) Workshop on Mathematics for Social Justice. Specifically, we use the aforementioned example from advanced statistics as a starting point for our explorations. We begin by providing a transcript of the conversation.

The sample conversation provided below was edited for clarity and to meet the editorial requirements of the article. We label the participants α, β, γ and so forth to maintain the anonymity of the participants. This conversation will allow readers to follow the key ideas that were discussed. The reflections provided at the end of the conversation are meant to provide a more in-depth understanding of participants' perspectives and help make sense of any nuances.

Participant α: A new innovation that departments are implementing is incorporating issues of social justice across their undergraduate curriculum. In fact, some colleges across the country are adding or modifying their mission statements to ensure exposure to issues of responsible leadership and democratic citizenship. I wonder what sort of contributions mathematicians might make here.

Participant β: I have been thinking about introducing social justice into my intermediate statistics class. One particular study that seemed relevant was Radelet's [25] discussion about the relationship between the race of a defendant and the statistics behind whether or not a defendant who was convicted of homicide would receive the death penalty. The study looked at cases from Florida in 1976 and 1977, but it can be a bit misleading (see table 1, a two-way table adapted from [18]):

α: How so?

β: In this table, it seems that Caucasians are more likely to be sentenced to death after being convicted of murder.

Participant γ: Okay, but before we talk about the details of this table, we should probably unpack the topic more so that we can ensure that students are able to fully understand the complexity of issues involved when it comes to race and U.S. law and public policy.

Participant δ: That's going to take a lot of time. I know that it is important for students to have a sufficient background on the topic, but as an instructor, I also have an obligation to present all of the material described by the course catalog. Many of my students will be taking actuarial exams and there is a set syllabus of topics that are intended to cover parts of the exam.

γ: But in order to have a productive discussion, we need to make sure that students have a more complete understanding of the history and terminology they are using. It is important for students to know that the words we choose can reflect different positions. Consider the difference between describing a group as illegal immigrants or undocumented persons, for example. Using different language can have a significant effect on the resulting conversation.

δ: But how is that going to help my students with their actuarial exams?

α: That's a good point γ and a good question δ. Let me take this a step further. What do you want your students to accomplish beyond your course? For example, do you want them to analyze the fairness problems in lending or look at the history that shows us how African Americans and other minorities are less likely to obtain home loans? A student could come into a regular actuarial mathematics class and leave with a different world view. Are we simply purveyors of technical knowledge? What is our role as teachers to ensure that our students are ultimately going to go on and change the world? Am I simply someone who transfers information?

β: I think I understand and I am sympathetic to your ideas, but is this really the purview of us as instructors? I'm not sure how comfortable I am in presenting my own viewpoints in a mathematics classroom.

γ: We should then challenge the assumption that mathematics is neutral. I know that we are not used to debating our answers or examining the social consequences of our solutions, but such courses are a great starting point for students to consider whether they have a moral responsibility for the applications of their work and research.

β: Well, one thing that I have wanted to try was flipping my classroom. It seems that we would really need class time for a careful discussion of these sensitive issues, but other more formulaic material might allow students to be investigators outside of class. I'm going to have to think about this some more.

δ: One concern that I have about teaching a class focused on issues of social justice is really a question: How do I make sure that all of my students feel comfortable? After all, there are a number of different political beliefs in relation to these ideas and topics.

α: The main thing that is important is that students are having a dialogue, especially if the classroom is flipped. I believe, as an instructor, your responsibility is to create an open environment where students can share their views. It is not necessarily about what conclusions are drawn

Table 2. A more complete picture

Race of Defendant	Race of Victim	Death Penalty	Lesser Sentence
Caucasian	Caucasian	19	132
	African American	0	9
African American	Caucasian	11	52
	African American	6	97

by the end of a class session, if any. If the arguments presented in class are strong, they will hold up over time. But we have to be sure to emphasize to students that they should use evidence to substantiate their ideas and not just rote methods. It's okay to disagree but it is not okay to lack civility in the discussion.

δ: How would I keep my international students engaged in a subject like racial profiling in the U.S.? Will they understand what this means?

α: We do not have to talk about issues of injustice from an entirely U.S. perspective. For example, we can use Social Progress Index measurements to look at social challenges in different countries, or see if there are similar instances of racial profiling across different countries.

β: Yes. But can we go back to the table for a moment? The situation is actually a bit more nuanced than what I initially described. Let us add an additional category: the race of the murder victim Table 2. This might lead to some different conclusions but a more complete picture.

β: The authors of this table state that factors other than the race of the victim might give similar results, but this is still a good lesson for students to see, and it provides more nuance.

δ: I found it interesting that the authors of the textbook [18] and this study highlighted this point.

γ: Thanks for showing us this example. I might try to use it in my statistics class next semester but more discussions need to be held about the implications of the table and its meanings.

δ: Thanks for sharing your ideas. This conversation has been really beneficial to me. From this short interaction, I have learned that I should not be too easily discouraged about incorporating social justice issues in my classes. I learned the value of starting small and framing our ideas. This conversation helped me to see the bigger picture, and also understand that this is really an ongoing conversation where I am not expected to have all the answers at once.

α: Yes, I agree. I have learned that we are all in very different places when it comes to this work. Continuing these conversations is important to developing our practices, but reflecting on our discussion is also very cruicial. I might suggest that in our next conversation we discuss the role of race in examples like these. Dr. Danny Martin, in an article on race and mathematics [21], talks deeply about the role of race in mathematics teaching and learning. I think we should engage some of his work to help us make sense of this example more and how it may affect different learners.

γ: I like that idea. Could you send us some of what you're thinking? We can set up a time to discuss our reflections and other applications again. I am okay with meeting virtually if we cannot find a time to meet in person.

In this excerpt from the conversation, some participants felt that time should be spent on providing context to social justice issues whereas others felt this should not be the main focus of a mathematics course. By developing practices to talk through multiple and concurrent questions about the example, the faculty members found ways to build upon their own knowledge in a CoP. The topic surrounded a common issue of injustice related to racial differences in statistics and it allowed the participants to think about the applications of this knowledge in practice. Importantly, as a result of the differences that surfaced in this conversation, participants noted that more conversations were needed.

4. MATHEMATICAL INTERPRETATIONS OF INJUSTICE

We now take a moment to examine more carefully both tables and their presumptive statistical implications from an applied mathematical point of view. We provide this analysis to help frame how faculty might approach discussions about the implications of critical conversations. Reviewing this information offers an opportunity to explore different interpretations of injustice that might surface during critical conversations on social justice in mathematics.

In Table 1, the percentage of Caucasian defendants sentenced to death was 19/160, or around 11.9%, and the percentage of African American defendants sentenced to death was 17/166, or around 10.2%. This computation suggestions that, in general, Caucasian defendants are more likely to be sentenced to death than African American defendants. However, in Table 2, the results suggests a different conclusion.

Specifically, Table 2 suggests that the death penalty is less likely to be imposed if the race of the victim was African American, regardless of the race of the defendant. For Caucasian defendants, the percentage sentenced to death changed from about 12.6% to 0.0% when the race of the victim changed from Caucasian to African American. Similarly, for African American defendants, the percentage sentenced to death changed from about 17.5% to about 5.8%. The phenomenon noted in these tables is known as Simpson's Paradox.

Simpson's Paradox [18], also known as the Yule–Simpson effect, occurs when frequency data is incorrectly used for causal interpretations. The paradox disappears when causal relations are considered and applied during interpretation. Understanding this paradox is important to making sense of both the mathematical implications and the social contexts that should be highlighted during critical conversations on social justice in mathematics. For example, the defendants were those convicted of homicide, but it is not clear how the pool of defendants in the table were chosen. There may have been other biases throughout the rest of the judicial process that significantly influenced the numbers in each of these tables. Importantly, these biases should be noticed and engaged within critical conversations.

In this specific example, it may be necessary to understand complex histories of race and the justice system, or take note of racial legislations embedded within U.S. laws and public policies (e.g., the Jim Crow laws, Chinese Exclusion Act, Civil Rights Laws). These laws have been specifically biased against African Americans and other marginalized, non-majoritarian (i.e., non-White) groups. Discussing the long-standing effects of factors of this nature should be viewed as a crucial aspect to framing and making sense of the goals of critical conversations.

Reflecting on the implications of critical conversations and participant reflections allow us to relate social issues to specific ideas about mathematics and its effects on our teaching practices. In doing so, faculty more naturally develop their professional practices by engaging in difficult discussions to identify barriers to implementation with their colleagues. As these practices are more routinely engaged, the potential for growth and the development of critical practices related to teaching mathematics for social justice are more clearly realized. In the next section, we present reflections from the previous conversation to understand its impact on individual participants.

5. PARTICIPANT REFLECTIONS

From the point of view of δ: As a novice instructor in the subject of mathematics for social justice, it was a great opportunity to meet and exchange ideas with instructors who are either new or seasoned in the field. The workshop encouraged me to start a similar dialogue on my campus. When I went back that summer, I discussed some of the subjects that came up in the ACS/FSU workshop conversations with my colleagues on campus (the workshop was only offered to mathematics and mathematics education faculty). On my campus, I was able to involve faculty members from other departments such as sociology, history, communications, and the sciences engage in conversations about how mathematics applies to social justice. Having small group discussions in an interdisciplinary setting helped me see the interest in integrating issues of social justice into mathematics and it allowed me to get ideas from different sources. The workshop provided us with the initial steps and helped me create a platform on our campus to continue these conversations. In our departmental discussions, we realized that it is more difficult to integrate such subjects in the context of upper-level mathematics courses. In those cases, our initial step was to include an assignment of writing a mathematical autobiography where students needed to describe their personal journey of learning mathematics and whether race/ethnicity/gender/socioeconomic status affected their success in mathematics. We hope that with such assignments, we can at least create a question mark in students' minds to realize the inequalities in mathematics learning. The idea of asking students to write a mathematical autobiography was suggested by α as one practical way to get students thinking about their own identities. The statement that was adapted from the syllabus was as follows:

> You will write a short paper reflecting on your prior experiences in mathematics. The prompt for this reflection paper is as follows: How have your race-ethnicity, socioeconomic status, gender, language capabilities, and other connected factors impacted your educational opportunities, outcomes, and others' expectations of you? The precise format for the paper is open; however, all submissions should be well written and answer each component of the prompt in full detail.

With this and other assignments, our ultimate goal is to design courses that will encourage students to think and care about the stories behind the numbers.

From the point of view of β: Before starting this conversation, I was excited at the prospect of incorporating issues of justice and inequality into my classroom. I thought that presenting some examples highlighting different races and ethnicities would be enough to engage student interest. What I learned was that these issues run much deeper. In order to accurately present and interpret data, students need some context to set the stage for such a discussion. This might be good justification for using a flipped model to create some additional class time to foster these discussions. I think now I have a better sense of the structure of how social justice issues can be addressed in an upper-level setting. Sometime in the first week or two I hope to set aside a day to focus on "context" and then follow this up in a second class with a case study. My hope for this case study would be to give students a chance to revisit statistical topics they might have seen in a previous course, but also view them through a new lens. The hope would be that students would continue this critical analysis throughout the course and there might be a few specific spots where critical in-class conversations would be highlighted. One specific change that I have made after the workshop was to include a first-day affirmation exercise that asks students to (privately) write down values important to them and then select their most important value from this list; this idea is based on an article from *Science* by Cohen [4]. Thus, I have found value in using resources from others to help me build my own practices.

From the point of view of α: As someone who has read a considerable amount of literature on critical theory and social justice pedagogy, these conversations were difficult at times. They were difficult mostly because of my realization, or lack thereof, of the definitions and applications of social justice that novice instructors may be interested in due to their ease of application. At times, I find that these applications considered by novice instructors are in direct opposition to the very foundations of the critical theories from which they are founded. It was, and still is, critically important to me that those who want to engage issues of social justice in their work continue to do the tough work of *reading* about long histories of injustice and engage in tough discussions about race and ethnicity, class, gender, and their intersections, to name only a few. Often, teachers and scholars silo themselves and began to take on new practices in the name of "social justice" but engage in "low ceiling" tasks to increase their comfort. Social justice work is not meant to be comfortable; by not taking the time to "do the work" (reading, collaborating, and understanding), our practices can have deleterious effects on the overarching purpose of teaching for social justice, and it can deeply impact our students. Still, I have learned the important lesson that it is always necessary to consider the fact that we all are in very different places when it comes to understanding *critical* issues of justice, and that we all must start somewhere—none of us were born with this knowledge. As a result, I have found that critical conversations are a vital pathway to helping myself and others make better sense of teaching mathematics for social justice, and I try to find ways to bridge the gap between theory, knowledge, and our classroom applications.

6. A FRAMEWORK FOR CRITICAL CONVERSATIONS

To present some components of the practices used for starting conversations on teaching mathematics for social justice, we reflected upon our own and other faculty experiences at the ACS Workshop on Social Justice in Undergraduate Mathematics. We use these to develop a framework. We found that developing a base cycle for critical conversations

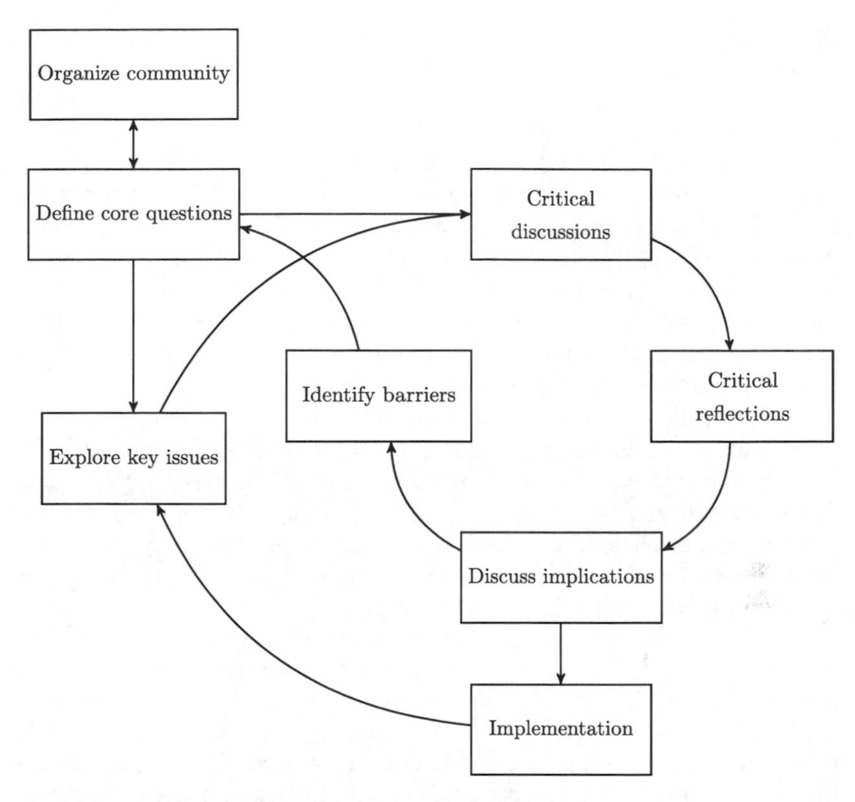

Figure 1. A sample process for critical conversations.

would be a useful start for developing a framework for a community practice as a means to generate more regular, and iterative, discussions.

Namely, we found that critical conversations can not effectively occur over a single interaction. Instead, they require multiple interactions to develop the CoP, and at times require differential foci during and within their development [1]. The base components of a critical conversation are given: (i) organize the community; (ii) Define core questions; (iii) Explore key issues; (iv) Critical discussion; (v) Critical reflections; (vi) Discuss implications; and (vii) Implementation. Each component has a generative purpose toward moving the conversation forward, and it importantly frames the actual discussion of any topic as only one part of the larger critical conversation.

Our framework presents a sample process for the development of critical conversations based on developing practices for teaching mathematics for social justice (see Figure 1). In our reviews, we found three common themes across participant's reflections and in the notes generated from participating community members. These themes focused on:

(i) framing and reflecting on the critical conversation; (ii) exploring implications and content connections; and (iii) identifying barriers. In general, participants recognized that there was a gap between this process and the content and pedagogical implications.

Framing and reflecting on the critical conversation. Framing a critical conversation begins with *organizing the community*. This group of participants, for example, may be considered as a CoP focused on a specific outcome or some specific goals. This initial process of organization will help to identify a community of practitioners [20, 30] who will engage in the critical conversation, and it provides an opportunity for organizers and participants to consider other community members who might be invited to join in on the conversation.

As a result of this initial organization, the group should begin by *exploring key issues* with the important note to start small and gradually develop and add content for future conversations. Exploring key issues, as noted by the CoP framework, can happen more naturally and result as a direct approach to the critical conversation. Alternatively, these key issues might be used to *define core questions* that can help the group structure and frame specific points of interest.

In the development of core questions, the group should consider the overall goals of their conversations. Will the group focus on understanding one specific idea or will they concentrate on multiple ideas and their implications for practice? Although seemingly related, the answers to these questions can have vastly different implications for the structure of the conversation and participant reflections. In both instances, however, exploring key issues develops into the next step in the process, the *critical discussion*.

Critical discussions, even within the same group of participants, will likely range in function, practice, and mechanics (e.g., time spent on topics). Instead of focusing on the specific mechanics of holding a critical discussion, we focus on some of the factors that will help to generate some of the boundaries that might be useful for the group. In the sample conversation presented, the focus of the discussion was related to a specific topic: integrating social justice issues into an advanced statistics course for undergraduates. Key considerations for this group included the following features: (a) the function: how will the discussion serve our collective and individual interests? (b) the practice: how can the discussion inform our collective and individual practices? and (c) the mechanics: what is the structure (or agenda) for the discussion and how much time will we spend on the various topics? Assessing these components prior to entering a critical discussion can help prepare the group for the transition into a period of critical reflection.

Critical reflections should occur throughout the larger practice of a critical conversation. However, setting aside specific time, generally after a discussion, will help the group to reflect as a group and allow participants to reflect individually. Some important questions to potentially ask during these scheduled reflection periods are: How did you feel before the group started the discussion? How did the discussion inform attitudes and beliefs about the topic of discussion? What did I learn from this conversation? How was this learning related to the different perspectives of others? Answers to these questions are likely to come easy when reflecting as a group; discussing the individual impacts of the conversations is a useful activity after individual reflections.

Exploring implications and content connections. Each participant in the sample critical conversation held different ideas about what was important when initiating the conversation. Furthermore, these differences carried through until after the discussion as participants worked to understand more about the conversation's implications. Once an initial conversation had been held, the group continued by exploring the implications of their discussions in multiple ways based on the outcomes of their reflections.

Additional reflections should occur within the larger group as participants *discuss implications* of their conversations. For example, in the sample discussion presented, additional individual reflections occurred after the larger group's reflections and they help to identify the various implications for specific practices based on participant's perspectives. However, your group might decide that this process should be reversed. In doing so, participants should come to some conclusions about how these reflections can be engaged in future discussions and how they will use them to discuss the implications of their conversations on their practices and as a means to identify any barriers.

Given the topic of the sample critical conversation presented on teaching mathematics for social justice, the primary implication of the group's discussions focused on content connections. Participants wanted to understand how the discussion would translate into their classroom practices. Researchers have examined similar barriers associated with negotiating mathematical goals in light of developing social justice curricula [2], it is necessary to acknowledge that even a seemingly straightforward mathematics problem with social justice connections might be difficult to understand.

In undergraduate mathematics settings, engaging social justice will likely be a new experience for students and instructors and thus can be a frightening or challenging experience. To help assuage these concerns, we suggest participants of the critical conversation take some additional time to explore any readings of relevant materials and integrate these materials into their discussions with other faculty participants. Using the CoP framework, we also suggest that faculty engage in the exploration of content connections as a group. These various professional practices can be described in the process that frames the lower half of the cyclical framework for critical conversations.

Identifying barriers. In light of the barriers highlighted in the sample conversation, it is important to understand not only the impact of the conversation on participants but also on some of the more practical aspects related to practice. Once the conversation is finished, participants enter into a critical period of reflection and discussion of implications. These implications might be used as a first step toward redesigning mathematics courses or integrating issues of social justice into the curriculum. However, identifying barriers as they relate to the goals of implementation is a crucial step in the overall process of developing critical conversations.

For example, discussing such problems in a small-group environment with other faculty members before using them in the classroom will be helpful in reflecting on the value and potential of teaching mathematics for social justice. Additionally, running interdisciplinary discussion groups with faculty from other departments on campus can provide new avenues for discussing efficient ways to implement mathematics into other courses across campus, and in particular those courses that have a social justice component. Currently, the annual MathFest, run by the Mathematical Association of America, and the Joint Mathematics Meetings offer special topic sessions on mathematics for social justice. These sessions aid faculty in thinking more practically about the steps related to teaching mathematics for social justice.

Engaging in similar discussions and communities of practice can help to surface specific institutional or departmental barriers to effective implementation. These barriers should be used specifically to explore a new set of key issues for the group, and ultimately begin the cyclical process inherent in critical conversations. Specifically, as new barriers are identified, they become a specific topic of focus for either the larger group or for smaller, topic-specific groups to restart and begin this process of critical conversations once again.

7. CONCLUSIONS AND IMPLICATIONS

In this article, we presented details about a faculty workshop that served as a backdrop to the analysis of a critical conversation. During the workshop, faculty members of mathematics from various U.S. institutions engaged in conversations about integrating undergraduate mathematics and social justice. We highlighted some of the core literature related to teaching and learning mathematics for social justice, based on critical pedagogy, and we used critical theories to help define critical conversations.

We also provided reflections on a sample conversation and we presented a number of key considerations from different perspectives. These reflections helped us to outline a number of key issues and considerations that should be addressed when moving from theory and incorporating aspects of social justice in undergraduate mathematics classrooms. These various aspects were used to develop a framework that presents a sample process for the development of critical conversations.

We focused our framework on some of the practical (and cyclical) aspects of preparing for and holding critical conversations. Upon organizing a diverse CoP, participants begin the process of framing the discussion. In framing the discussion, community members explore key issues and decide if these issues will be developed more organically into a conversation or used to define core questions. In each of these pathways, the conversation is framed by the community of participants and the issues identified.

In particular, we show how these dialogues can expand the collective knowledge of different practices surrounding teaching and learning. To do so, we discuss critical conversations in two distinct ways: first, as a theoretically vital component to faculty's professional commitment to equity and social justice and, second, as a practical component to developing one's professional practices in this area over time. Together these two components frame the potential professional development work informed by critical conversations.

In this article, we utilize the term critical conversation to discuss practical forms of professional development that sit within the larger space of regular, professional interactions and work-related dialogue. We note that some everyday discussions might be better structured into what we have named critical conversations and that these conversations can help to build upon and develop faculty's teaching practices. Our discussions focus on the various components of critical conversations as they relate to integrating social justice issues into undergraduate mathematics courses.

We use the lens of everyday discourses, in the form of conversations, to explore how discussions about teaching might be transformed to intersect non-dominant perspectives into dominant ones, as well as help negotiate multiple perspectives on teaching mathematics for social justice. The term "critical" helps us to define, explore, and examine how faculty might deal with differential and/or opposing views within these discourses. To exemplify this process, we reflect on a conversation that took place at a professional meeting focused on teaching and learning mathematics for social justice.

Towards this end, we witnessed the power of critical conversations on social justice in mathematics to help faculty make sense of different attitudes and beliefs about social injustices and their historical effects. They allowed collaborators to relate them to mathematics teaching and learning. These conversations aided us in shifting our ideas through critical self-reflections and collaborative practices focused on different conceptions of social justice.

Critical conversations on social justice in mathematics also matter because they provide opportunities for self-reflection and collaboration in expanding perspectives on teaching and learning. These conversations present colleagues with opportunities to engage in difficult discussions about the presence of systems of injustice.

Our overall goal in this discussion was to present a starting point for individuals and departments hoping to integrate critical conversations at their home institutions, and to use these conversations to explore conceptions of social justice in teaching mathematics. We hope that novice instructors begin to engage in understanding social justice as a tool not only for teaching, but also for learning how to embody social justice in their classrooms and as a CoP. Through critical conversations, our daily practices can begin to move us collectively toward a more inclusive society.

ACKNOWLEDGEMENTS

The authors would like to thank and acknowledge the organizers and participants of the ACS/FSU Workshop on Social Justice in Mathematics. The authors would like to especially thank Bonnie J. Shulman and Paul H. Kvam for helping to foster critical conversations in our small-group discussions. Finally, the authors would like to thank the referees for their helpful comments.

FUNDING

We would also like to thank the ACS and FSU for theirfinancial support of the workshop.

REFERENCES

1. Allan, B. 2008. Knowledge creation within a community of practice. Unpublished manuscript.

2. Bartell, T. G. 2013. Learning to teach mathematics for social justice: Negotiating social justice and mathematical goals. *Journal for Research in Mathematics Education*. 44(1): 129–163.
3. Bell, D. A. 1992. *Faces at the Bottom of the Well: The Permanence of Racism*. New York, NY: Basic Books.
4. Cohen, G. L., J. Garcia, V. Purdie-Vaugns, N. Apfel, and P. Brzustoski. 2009. Recursive processes in self-affirmation: Intervening to close the minority achievement gap. *Science*, 324: 400–403.
5. Bell, D. A. 1987. *And We Are Not Saved: The Elusive Quest for Racial Justice*. New York, NY: Basic Books.
6. Durkheim, E. 1982. *The Rules of Sociological Method*. New York, NY: Simon and Schuster (The Free Press).
7. Frankenstein, M. 1987. Critical mathematics education: An application of Paulo Freire's epistemology. In I. Shor (Ed.), *Freire for the Classroom: A Sourcebook for Liberatory Teaching*, pp. 180–210. Portsmouth, NH: Boynton/Cook.
8. Freire, P. 1970. The adult literacy process as cultural action for freedom. *Harvard Educational Review*. 40: 205–225.
9. Freire, P. 1972. *Pedagogy of the Oppressed*. New York, NY: Herder and Herder.
10. Freire, P. 1985. *The Politics of Education: Culture, Power, and Liberation*. Hadley, MA: Bergin & Garvey.
11. Freire, P. 1994. *Pedagogy of Hope: Reliving Pedagogy of the Oppressed (R. B. Barr, Trans.)*. New York, NY: Continuum.
12. Gewirtz, S. 1998. Conceptualizing social justice in education: Mapping the territory. *Journal of Education Policy*. 13(4): 469–484.
13. Giroux, H. A. 2007. Democracy, education, and the politics of critical pedagogy. In P. McLaren and J. L. Kincheloe (Eds), *Critical Pedagogy: Where Are We Now?* pp. 1–5. New York, NY: Peter Lang.
14. Gutstein, E. 2003. Teaching and learning mathematics for social justice in an urban, Latino school. *Journal for Research in Mathematics Education*. 34: 37–73.
15. Gutstein, E. 2006. *Reading and Writing the World with Mathematics*. New York, NY: Routledge.
16. Gutstein, E. and B. Peterson. (Eds). 2013. *Rethinking Mathematics: Teaching Social Justice by the Numbers, second edition*. Milwaukee, WI: Rethinking Schools.
17. Kokka, K. 2015. Addressing dilemmas of social justice mathematics instruction through collaboration of students, educators, and researchers. *Educational Considerations*. 43(1): 13–21.
18. Kvam, P., B. Vidakovic, and S. Kim. 2015. Categorical Data. In P. H. Kvam and B. Vidakovi (Eds.), *Nonparametric Statistics with Applications in Science and Engineering*, pp. 155–182. New York, NY: John Wiley & Sons, Inc.
19. Larnell, G. V., E. C. Bullock, and C. C. Jett. 2016. Rethinking teaching and learning mathematics for social justice from a critical race perspective. *Journal of Education*. 196(1): 19–29.

20. Lave, J. and E. Wenger. 1991. *Situated Learning: Legitimate Peripheral Participation*. Cambridge, UK: Cambridge University Press.
21. Martin, D. B. 2009. Researching race in mathematics education. *Teachers College Record*. 111(2): 295–338.
22. Martin, J. 1996. *The Dialectical Imagination: A History of the Frankfurt School and the Institute of Social Research, 1923 - 1950*. Berkeley, CA: University of California Press.
23. Moses, R. P. and C. E. Cobb. 2001. *Radical Equations: Math Literacy and Civil Rights*. Boston, MA: Beacon Press.
24. North, C. E. 2006. More than words? Delving into the substantive meanings of "social justice" in education. *Review of Educational Research*. 76(4): 507–535.
25. Radelet, M. 1981. Racial characteristics and the imposition of the death penalty. *American Sociological Review*. 46(6): 918–927.
26. Schoenfeld, A. 2004. The math wars. *Educational Policy*. 18: 253–286.
27. Skovsmose, O. 1994. Towards a critical mathematics education. *Educational Studies in Mathematics*. 27: 35–57.
28. Stinson, D. W., C. R. Bidwell, and G. C. Powell. 2012. Critical pedagogy and teaching mathematics for social justice. *The International Journal of Critical Pedagogy*. 4(1): 76–94.
29. Tate, W. F. 1997. Critical race theory and education: History, theory, and implications. *Review of Research in Education*, 22: 195–247.
30. Wenger, E. 1998. *Communities of Practice: Learning, Meaning, and Identity*. Cambridge, UK: Cambridge University Press.

Index

Note: *Italic* page numbers refer to figures

9781032058252